LEARN & EXPLORE

New Views of the Solar System

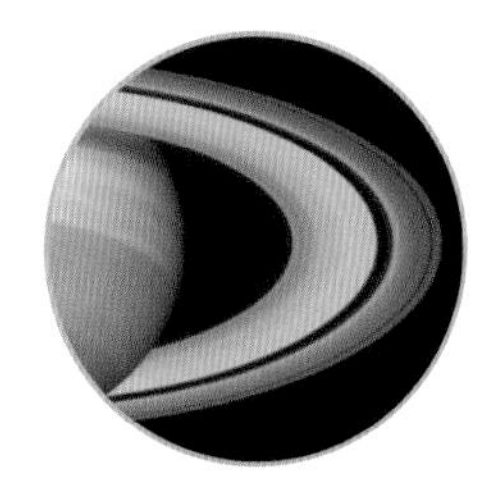

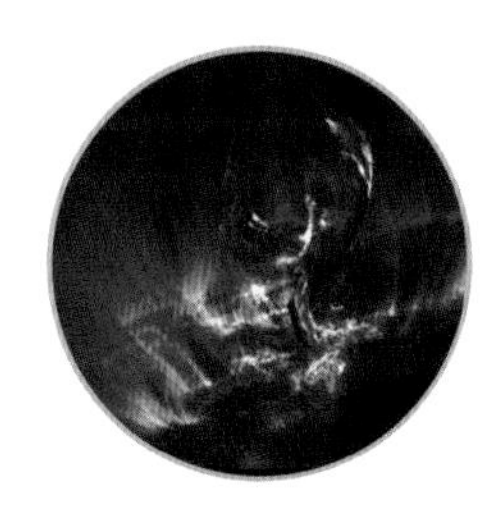

Compton's by Britannica®

CHICAGO LONDON NEW DELHI PARIS SEOUL SYDNEY TAIPEI TOKYO

Learn & Explore series
New Views of the Solar System
Compton's by Britannica

Library of Congress Control Number: 2007908762
International Standard Book Number: 978-1-59339-890-3

Printed in China

Britannica may be accessed at http://www.britannica.com on the Internet.

Front cover (top): Magellan spacecraft, NASA/JPL
Front and back cover (center): illustration of the eight planets, NASA/Lunar and Planetary Laboratory
Front cover (bottom): the sun, SOHO/ESA/NASA
Back cover (top): NASA
Back cover and title page (left to right): Saturn, NASA/The Hubble Heritage Team; solar prominence, TRACE/NASA; Saturn's moon Rhea, NASA; artist's concept of the Mars Exploration Rover, NASA/JPL/California Institute of Technology

EDITOR'S PREFACE

"The eight chief planets, in the order of their distance from the sun, are Mercury, Venus, Earth, Mars, Jupiter, Saturn, Uranus, and Neptune." That quote is taken from the 1929 edition of *Compton's Encyclopedia*, prior to the discovery of Pluto. U.S. astronomer Percival Lowell had predicted the existence of a planet beyond the orbit of Neptune. This initiated the search that ended in Pluto's discovery by Clyde Tombaugh at Lowell Observatory at Flagstaff, Ariz., on Feb. 18, 1930. For the next 76 years Pluto was considered our solar system's ninth planet.

There is a popular saying that goes, "everything old is new again." In August 2006 the International Astronomical Union, the organization that approves the names of astronomical objects, made the controversial decision to redefine the term "planet." The new definition in effect stripped Pluto of its designation as the solar system's ninth planet, making the 1929 *Compton's* quote cited above valid once again. As of 2006 Pluto was to be known as a dwarf planet.

This book takes a look at our new view of the solar system with articles on the eight planets presented in order of their distance from the sun. In addition there are articles on planet, dwarf planet, Pluto, and others, that will help to illustrate this "new view." The articles are presented here with an equal concentration on text, photos, art, and tables. The first edition of *Compton's* appeared in 1922. It was the first pictured encyclopedia—that is, the first to use photographs and drawings on the same pages with the text they illustrated. This book holds true to the *Compton's* tradition "to inspire ambition, to stimulate the imagination, to provide the inquiring mind with accurate information told in an interesting style."

TABLE OF CONTENTS

NEW VIEWS OF THE SOLAR SYSTEM

This introduction was contributed by Dr. Clark R. Chapman, Senior Scientist, Department of Space Studies, Southwest Research Institute, Boulder, Colorado, and author of Cosmic Catastrophes *and others. Dr. Chapman is a past President of Commission 15 of the International Astronomical Union (which deals with physical properties of comets and asteroids) and a past Chair of the Division for Planetary Sciences of the American Astronomical Society. Unless otherwise noted, all articles in this book have been critically reviewed by Dr. Chapman.*

INTRODUCTION

Until 2006, there were nine planets, ranging from Mercury out to Pluto, which usually lies beyond Neptune. In August 2006, international astronomers, after much argument, eventually voted that Pluto was to be reclassified as a "dwarf planet." Ignoring conventional usage, they also voted that a dwarf planet is *not* a planet. For decades following its discovery in 1930, Pluto was thought to be roughly Earth-sized. Since the late 1970s, when Pluto's moon Charon was discovered, it has been realized that Pluto has just 1/500th the mass of Earth.

Two other astronomical developments forced reconsideration of the definition of "planet." In 1995, the first planet orbiting another star was detected. At least 200 extrasolar planets have since been found, and discoveries continue. Despite search biases favoring large planets close to the stars they orbit, most planetary systems differ greatly from the solar system. Many have large, Jupiter-sized planets moving very close to their stars in highly elongated orbits, unlike the more distant, nearly circular orbits of Jupiter and Saturn.

In 1992, a small body was discovered beyond Pluto. Since then, more than 1,000 of these so-called Kuiper Belt Objects (KBOs) have been found, some quite large. In 2005, an object now named Eris was found, which is somewhat larger than Pluto. Is it the "tenth planet"? Other nearly Pluto-sized KBOs have been sighted, and astronomers expect that others may exist that are even larger than Eris.

The International Astronomical Union (IAU), founded in 1919, has about 9,000 members from 85 countries. Traditionally the IAU has defined official names for asteroids, craters on planets, and so on. Although the IAU had never previously defined the word "planet," it decided to do so. After failure of its 19-member panel to reach consensus, the IAU appointed a new group of seven astronomers and scholars to define "planet." During a meeting in Paris in June 2006, the group agreed that a planet must be large enough to be approximately spherical because its gravity overcomes its material strength and crushes any large departures from sphericity. That definition, with some controversial addenda, was announced at the beginning of the IAU's General Assembly (held every three years) in Prague. The definition would admit such diverse bodies as Eris, the largest asteroid (Ceres), and Pluto's moon (Charon) into planethood, but not our own moon, Titan, or other Mercury-sized moons. Soon there could be dozens of such "planets."

Many astronomers strongly objected. Physicists who study how bodies orbit each other argued that planethood should be defined by dynamical properties; after all, the original definition of planets as "wandering stars" was based on their motions. The planet-definition committee's recommendations were subsequently rejected. Now, a planet must not only be largely spherical, but it also must be massive enough for its gravity to have cleared its orbital neighborhood of smaller bodies, a theoretical concept. Objects large enough to be round but deemed too small to have cleared their zones are now to be called "dwarf planets." All other solar system bodies, except the sun itself and moons of other bodies, are to be called "small solar-system bodies" of one sort or another—those asteroids, comets, KBOs, etc., that fail the roundness test. The IAU decided not to deal with defining extrasolar planets.

Pluto is no longer considered a planet by the IAU. Aspects of the new definitions are opposed by many planetary scientists, including planetary geologists, planetary atmospheric scientists, astrobiologists, cosmochemists, and others who study planets but were not at the astronomers' meeting in Prague. (The few hundred who remained for the final votes were mostly stellar and galactic astronomers, not planetary astronomers.) Important though it is that things have names and are grouped into categories, the continuing controversy over "What is a planet?" reflects cultural values; it is not science.

(Clockwise from left) SOHO/ESA/NASA; NASA/JPL; NASA; NASA/Space Telescope Science Institute; NASA

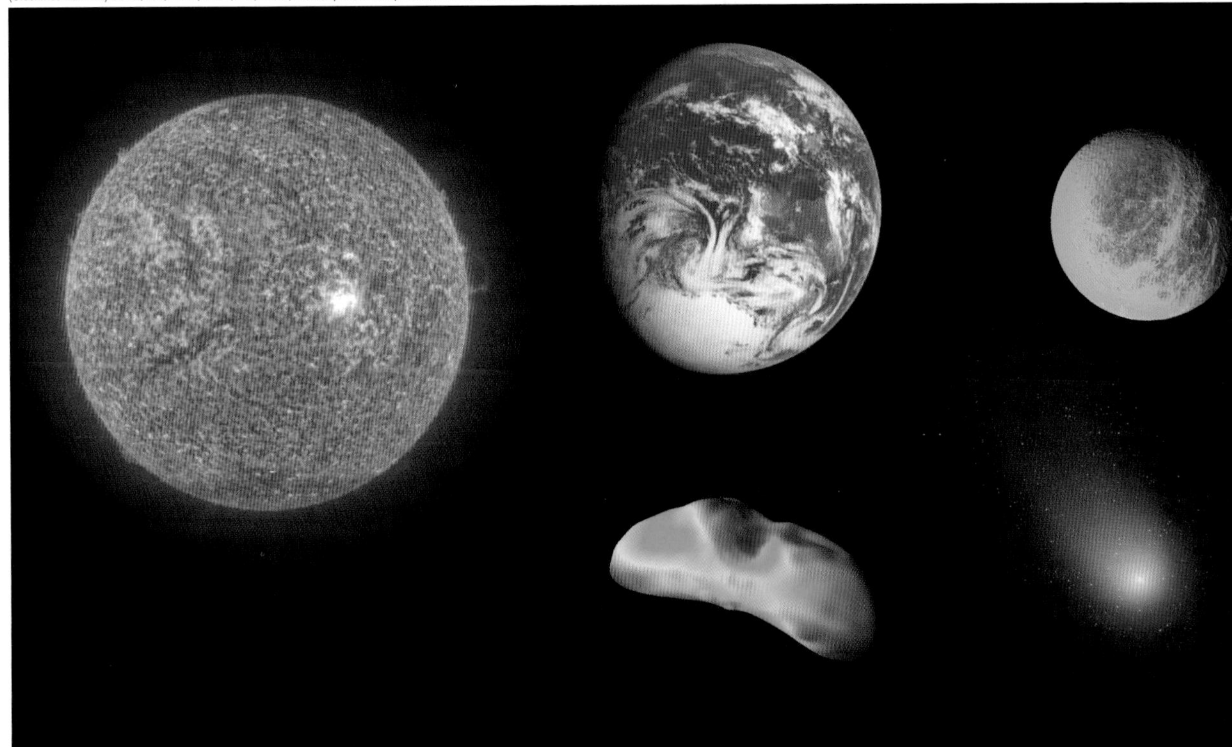

The solar system consists of the sun (left) and all the bodies that orbit it, including planets such as Earth (top center), satellites such as Saturn's moon Rhea (top right), comets such as NEAT (bottom right), and asteroids such as Eros (bottom center). The image of the sun shows a bright active region in its lower atmosphere, at right of center. It was taken in extreme ultraviolet light, with false color added in processing. In the image of Earth, the visible landmass is Australia. The moon Rhea is shown in false color to accentuate the wispy markings across its surface. The close-up view of Comet NEAT, officially called C/2001 Q4, shows its coma and the inner part of its tail. The image of Eros shows a model of the asteroid that was color coded to indicate the topography of its gravity at the surface. Objects on Eros would tend to move from red "uphill" areas to blue "downhill" areas.

SOLAR SYSTEM

As the sun rushes through space at a speed of roughly 150 miles (250 kilometers) per second, it takes many smaller objects along with it. These include the planets and dwarf planets; their moons; and small bodies such as asteroids, comets, and meteoroids. All these objects orbit, or revolve around, the sun. Together, the sun and all its smaller companions are known as the solar system. The solar system itself orbits the center of the Milky Way galaxy, completing one revolution about every 225 million years.

Earth is one of the larger bodies of the solar system. It is quite small, however, compared to the sun or the planet Jupiter, which are the largest members of the solar system. The solar system's smallest members are the microscopic particles of dust and the even smaller atoms and molecules of gas of the interplanetary medium. This dust and gas is very thinly scattered in the huge expanses between the planets and other bodies in the solar system.

THE SOLAR SYSTEM IN SPACE

Astronomers do not know exactly how far out the solar system extends. Earth orbits the sun at an average distance of about 93 million miles (150 million kilometers). Astronomers use this distance as a basic unit of length in describing the vast distances of the solar system. One astronomical unit (AU) is defined as the average distance between Earth and the sun.

There are eight planets in the solar system. Neptune, the outermost planet, orbits the sun from about 30 AU, or 2.8 billion miles (4.5 billion kilometers), away. Many comets have orbits that take them thousands of times farther out than Neptune. Most comets are thought to originate in the outermost parts of the solar system, the Kuiper belt and the much more distant Oort cloud. Each of these consists of countless small icy bodies that orbit the

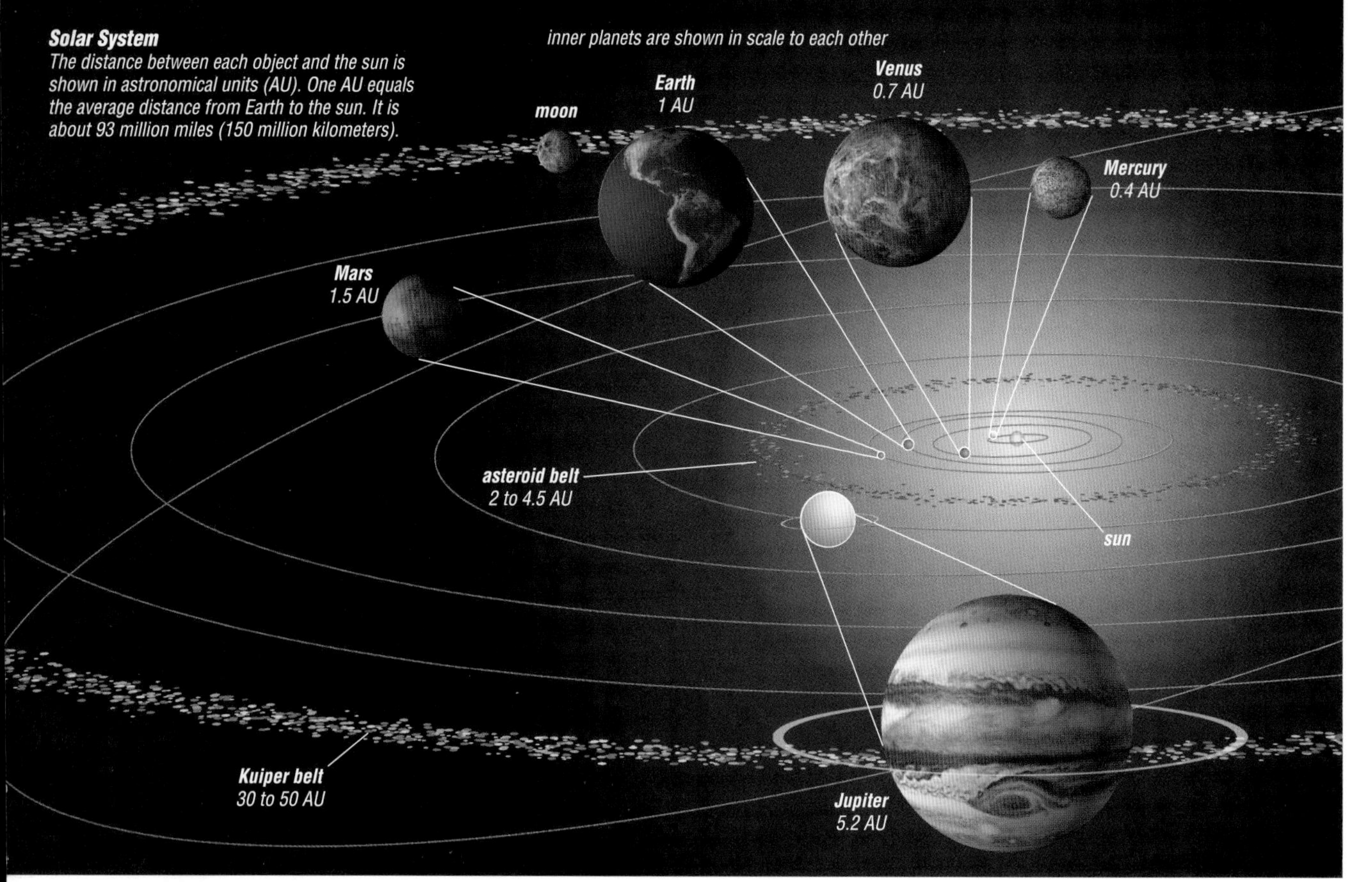

The solar system consists of the sun and all the objects that orbit it, including planets, dwarf planets, moons, and small bodies such as asteroids, comets, and the comet nuclei in the Kuiper belt and the Oort cloud. The drawing is not to scale overall. The representations of the Kuiper belt and the Oort cloud are simplified; the former is actually a doughnut-shaped zone, while the latter is a fairly spherical shell.

sun. The farthest reaches of the Oort cloud extend perhaps to 100,000 AU, or some 9.3 trillion miles (15 trillion kilometers), from the sun.

The solar system is, of course, not alone in space. The sun is a star like countless others, and other stars also have planets circling them (*see* Planet). The sun is part of the Milky Way galaxy, a huge group of stars swirling around in a pinwheel shape (*see* Astronomy). The galaxy contains hundreds of billions of stars. To measure the enormous distances in space, astronomers often use the light-year as a unit of length. One light-year is equal to the distance light travels in a vacuum in one year, about 5.88 trillion miles (9.46 trillion kilometers). The Milky Way galaxy is roughly 150,000 light-years across. The sun's nearest neighbor in the galaxy is the star Proxima Centauri (part of the triple-star system named Alpha Centauri). This "neighbor" lies some 4.3 light-years, or more than 25 million miles (40 trillion kilometers), away from the sun.

Outside the Milky Way galaxy there are billions more galaxies stretching out through space. Astronomers cannot see to the end of the universe, but they have detected galaxies and other objects that are several billion light-years away from the sun. Compared with such distances, the space that the solar system occupies seems tiny.

PARTS OF THE SOLAR SYSTEM

The sun is the central and dominant member of the solar system. Its gravitational force holds the other members in orbit and governs their motions. The largest members of the solar system after the sun are the planets and dwarf planets and their moons. The other natural bodies in the solar system are called small bodies. They include asteroids, meteoroids, comets, and the billions of icy objects in the Kuiper belt and Oort cloud.

The small bodies and the smaller moons can be quite irregularly shaped. The planets, the dwarf planets, and the larger moons are nearly spherical in shape. They are large enough so that their own gravity squeezes them into about the shape of a ball. The shapes of the planets and dwarf planets that rotate especially rapidly are distorted to various degrees. Instead of being perfect spheres, such bodies have some flattening at the poles, which makes them appear squashed.

Most objects in the solar system have elliptical, or oval-shaped, orbits around the sun. These objects include the planets, dwarf planets, asteroids, comets, and Kuiper belt objects. The planets orbit the sun in nearly circular orbits, while the small bodies tend to have much more eccentric, or elongated, orbits. The

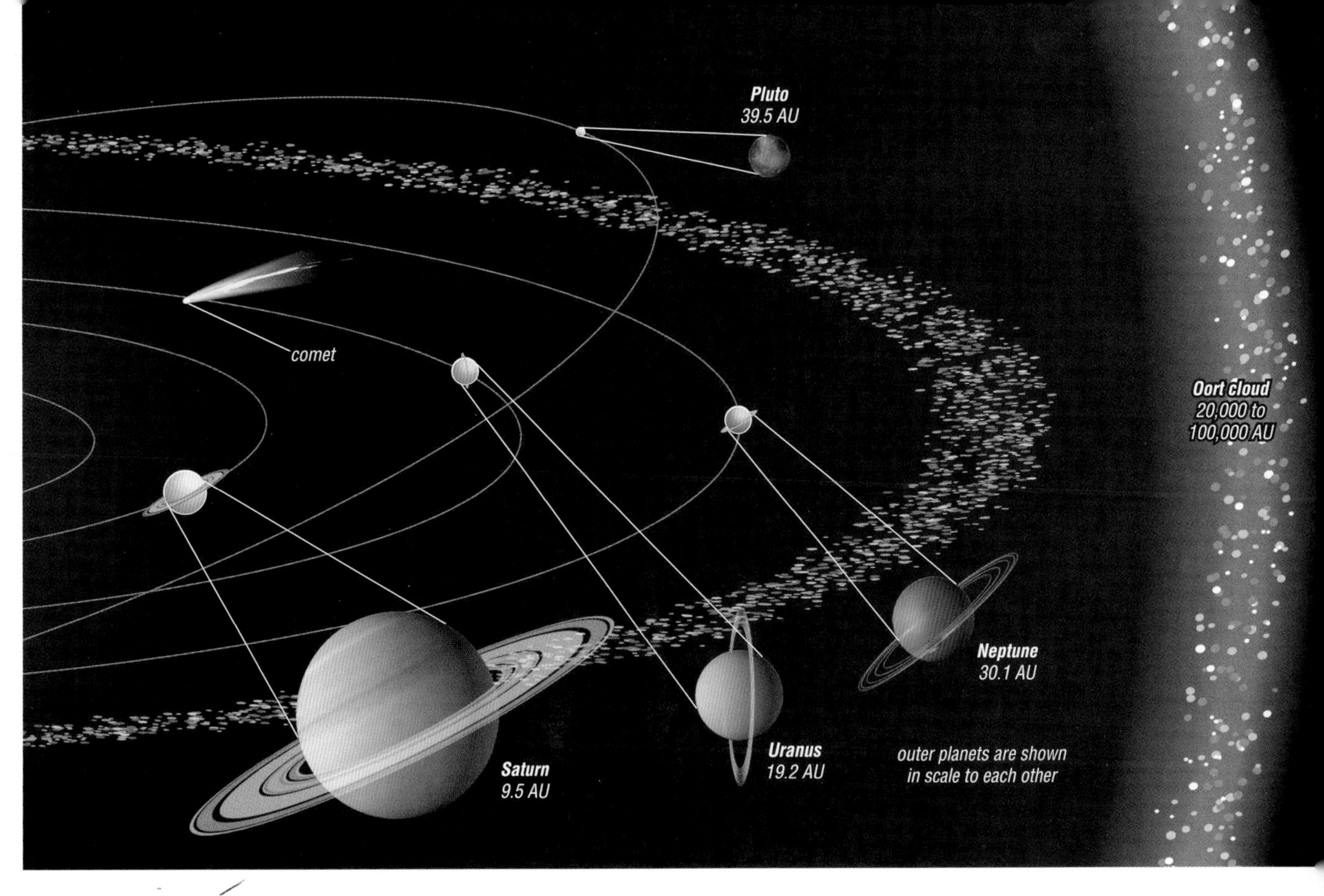

planets also orbit in very nearly the same plane, that of the sun's equator. The small bodies again differ, generally orbiting in planes that are more inclined, or tilted, relative to the plane of the sun's equator. Comets whose orbits take them very far from the sun tend to have especially eccentric and inclined orbits.

Many of those comets also orbit in a different direction than most other objects in the solar system. The sun rotates in a counterclockwise direction as viewed from a vantage point above Earth's North Pole. All the planets, dwarf planets, asteroids, and Kuiper belt objects and many comets orbit the sun in the same direction that the sun rotates. This is called prograde, or direct, motion. The comets with large orbits and the icy bodies of the Oort cloud are thought to be distributed randomly in all directions of the sky. Many of these objects orbit the sun in retrograde motion, or the direction opposite to that of the sun's rotation.

The Sun

The sun far outweighs all other components of the solar system combined. In fact, the sun contains more than 99 percent of the mass of the entire solar system. Nevertheless, the sun is a fairly average-sized star. From Earth it looks so much larger and brighter than other stars only because it is so much nearer to Earth than any other star. If the sun were much farther away, it would look pretty much like many other stars in the night sky. But if this were so, life as we know it could not exist on Earth. The sun provides nearly all the heat, light, and other forms of energy necessary for life on Earth. In fact, the sun provides the great majority of the energy of the solar system. (*See also* Sun.)

The Planets and Dwarf Planets and Their Moons

The largest and most massive members of the solar system after the sun are the planets. Even so, their combined mass is less than 0.2 percent of the total mass of the solar system, and Jupiter accounts for a very large share of that percentage. From nearest to farthest from the sun, the eight planets are Mercury, Venus, Earth, Mars, Jupiter, Saturn, Uranus, and Neptune.

Pluto had been considered the solar system's ninth planet from the time of its discovery in 1930 until 2006, when the International Astronomical Union (the organization that approves the names of astronomical objects for the scientific community) changed its designation. The organization created a new category of object called dwarf planet and made Pluto, Eris, and Ceres the first members of the group. Pluto and Eris are also considered Kuiper belt objects, and Ceres is also the largest asteroid. As their name suggests, dwarf planets are similar to the eight major planets but are smaller. (For a fuller discussion of the reclassification of Pluto and of the definitions of "planet" and "dwarf planet," *see* Planet.)

The eight planets can be divided into two groups, the inner planets and the outer planets, according to

NASA/Johnson Space Center/Earth Sciences and Image Analysis Laboratory

A luminous loop of the southern lights appears above Earth in an image taken from aboard NASA's space shuttle* Discovery*. Such auroras are caused by electrically charged particles of the solar wind colliding with gases in Earth's atmosphere.

their nearness to the sun and their physical properties. The four inner planets—Mercury, Venus, Earth, and Mars—are composed mostly of silicate rock and iron and other metals in varying proportions. They all have solid surfaces and are more than three times as dense as water. These rocky planets are also known as the terrestrial, or Earth-like, planets.

In sharp contrast, the four outer planets—Jupiter, Saturn, Uranus, and Neptune—have no solid surfaces. Jupiter and Saturn consist mainly of liquid and gaseous hydrogen and helium; Uranus and Neptune have melted ices and molten rock as well as hydrogen and helium. All the outer planets are less than twice as dense as water. In fact, Saturn's density is so low that it would float if put in water. The outer planets are also much larger than the inner planets, and they have deep gaseous atmospheres. Because of this, they are sometimes called the gas giants. Since Jupiter is the outstanding representative of this group, they are also known as the Jovian, or Jupiter-like, planets.

The eight planets are not distributed evenly in space. The four inner planets are much closer to each other than the four outer planets are to one another.

Six of the eight planets have smaller bodies—their natural satellites, or moons—circling them. All the outer planets have numerous moons: Jupiter and Saturn have more than 50 known moons each, Uranus has more than 25, and Neptune has more than 10. The inner planets have few or none: Mars has two moons,

The four inner planets are rocky worlds with solid surfaces. Porous rocks are strewn across a sandy slope in an image of the Martian surface captured by Spirit, one of NASA's Mars Exploration Rovers. The boulder in the foreground is about 16 inches (40 centimeters) high. It likely is made of basalt and, like earthly basalts, was originally formed by an active volcano. The approximately true-color image shows a rise dubbed Low Ridge in Gusev Crater.

NASA/JPL—Caltech/Cornell/NMMNH

NASA/JPL/University of Arizona

The four outer planets are huge worlds without solid surfaces. Each has many moons. The large moon Io, at center right, casts a shadow on the giant planet Jupiter in an image taken by NASA's Cassini spacecraft. Io is a bit larger than Earth's moon. Jupiter's striped patterns are bands in the clouds of its thick, stormy atmosphere. The enormous, long-lived storm called the Great Red Spot is visible at center left.

Earth of course has only one, and Venus and Mercury lack moons. Many Kuiper belt objects, including the dwarf planets Pluto, Eris, and Haumea also have moons, as do some asteroids.

The largest natural satellite in the solar system is Jupiter's moon Ganymede. Next in size are Saturn's moon Titan, Jupiter's Callisto and Io, Earth's moon, and Jupiter's Europa. Both Ganymede and Titan are larger than the planet Mercury. Earth's moon is so large with respect to Earth that the two bodies have sometimes been considered a double-planet system. The solar system's smallest moons, most of which orbit Jupiter and Saturn, are only a few miles in diameter.

Most of the solar system's larger moons, including Earth's, orbit their planet in the same direction in which the planets orbit the sun. A notable exception is Triton, which is Neptune's largest moon. It orbits in retrograde motion, as do many of the small, outer moons of the gas giants. Most of the solar system's moons also orbit their planet in the plane of the planet's equator. Again, Triton and many of the small, outer moons of the outer planets are exceptions, having highly inclined orbits. Moons that orbit in retrograde motion or that have inclined orbits or both are called irregular moons.

Saturn's spectacular rings are well known, but all the other outer planets also have systems of thin, flat rings. Each of the rings is composed of countless small pieces of matter orbiting the planet like tiny satellites. None of the inner planets has rings. (*See also* Planet; Dwarf Planet.)

Asteroids

Numerous rocky small bodies are called asteroids or minor planets. Their orbits lie, for the most part, in a doughnut-shaped zone between the orbits of Mars and Jupiter. This zone is known as the main asteroid belt. The asteroids are not distributed evenly in the main belt. Rather there are several gaps in their orbits, owing to the influence of Jupiter's gravitational force. The asteroids outside the main belt include the near-Earth asteroids, which come within at least about 28 million miles (45 million kilometers) of Earth's orbit. The orbits of some of these asteroids even cross Earth's orbit.

Ceres is the largest asteroid, with a diameter of roughly 585 miles (940 kilometers). The asteroids Pallas and Vesta each have a diameter greater than 300 miles (485 kilometers). Few asteroids, however, are larger than 100 miles (160 kilometers) across, and the numbers of asteroids increase dramatically at smaller sizes. It is estimated that millions of asteroids of boulder size exist in the solar system.

Astronomers think that asteroids are chunks of material left over from the process that created the inner planets. The huge pull of Jupiter's gravity prevented these rocky chunks from clumping together into a large planet. Many of the smaller asteroids are thought to be fragments caused by collisions between the larger asteroids. Some of these fragments collide with Earth as meteorites.

A child touches Ahnighito, one of the largest meteorites ever found, at the American Museum of Natural History in New York City. It is also called the Cape York meteorite. The 34-ton iron meteorite crashed into the ground in what is now Greenland, probably many thousands of years ago. It probably came from the core of an asteroid.

Jonathan Blair/Corbis

D. Nunuk/Photo Researchers, Inc.

A meteor streaks across the sky in a time-exposure photograph taken in British Columbia, Canada. It is part of the Perseid meteor shower, which occurs each summer in the Northern Hemisphere.

NASA/JPL

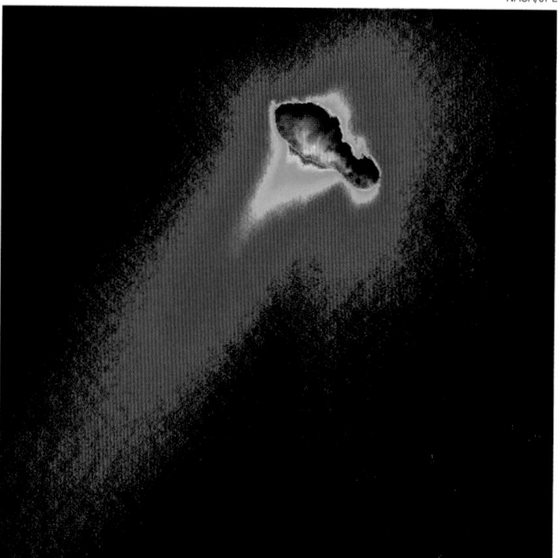

A false-color composite of images taken by the Deep Space 1 spacecraft shows Comet Borrelly's nucleus, dust jets, and coma (its hazy, dusty atmosphere). The nucleus, which appears gray, is about 5 miles (8 kilometers) long. The main jet of dust escaping from the nucleus extends to the bottom left. The comet's nucleus is the brightest part of the image. The other features have been color coded so that red indicates areas that are about a tenth as bright as the nucleus, blue a hundredth, and purple a thousandth.

The three main types of asteroids seem to be rich in organic compounds, stony materials, and iron and other metals, respectively. Some asteroids are thought to contain samples of the first materials to coalesce out of the great cloud of gas and dust from which the solar system itself is believed to have formed. (*See also* Asteroid.)

Meteoroids

Meteoroids are small chunks of rock, metal, or other material in interplanetary space. The vast majority of meteoroids are small fragments of asteroids. Other meteoroids are fragments from Earth's moon or Mars, while some may be rocky debris shed by comets.

When a meteoroid collides with Earth's atmosphere, it is usually vaporized by heat from friction with the air molecules. The bright streak of light that occurs while the particle vaporizes is called a meteor. Occasionally, a large chunk of rock and metal survives the journey to the ground. Such remnants are called meteorites. (*See also* Astronomy.)

Comets, the Kuiper Belt, and the Oort Cloud

At times, a fuzzy spot of light, perhaps with a tail streaming away from it, appears in the sky. Such appearances of the small icy bodies called comets are spectacular but infrequent. Comets are only easily visible from Earth when they pass close to the sun. Most comets that are detected from Earth are visible only with a telescope. Occasionally, one can be seen with the unaided eye, and several times a century a comet will appear that can be seen even in the daytime.

Just as the asteroids are the rocky remnants of the process that formed the inner planets, the comets are thought to be leftover icy material from the formation of the outermost planets, Uranus and Neptune. Comets contain particles of rock dust and organic compounds, water ice, and ices of various substances that are normally gases on Earth. As a comet approaches the sun, the ices turn to vapor, forming a hazy, gaseous atmosphere, or coma, around a core of solid particles, the nucleus. As the comet moves even closer to the sun, even more material is vaporized. Radiation and high-energy particles streaming out from the sun may push material away from the comet into one or more long, glowing tails, which point generally away from the sun. Comets either disintegrate completely, ending up as swarms of tiny particles, or continue along their orbital path. As a comet moves away from the sun, it loses its coma and tail. The only permanent part of the comet is its solid nucleus.

A comet may exist as a nucleus for thousands of years or more in the cold outer reaches of the solar system before its orbit takes it near the sun again. The billions of distant icy objects in the Kuiper belt and the Oort cloud are, in fact, comet nuclei. These objects are thought to have orbited as nuclei for eons, without ever having approached the sun. Sometimes the orbit of one of the objects in the Kuiper belt or the Oort cloud is disturbed by the gravity of another body so that the object is sent on a path that takes it closer to the sun. This process is thought to be the source of most comets. For instance, one way the orbits of Oort

Travel Time from Earth to Various Objects in the Solar System and Beyond*

Destination	Jet, 600 miles (965 kilometers) per hour	Rocket, 25,000 miles (40,250 kilometers) per hour	Ray of light, 186,282 miles (299,792 kilometers) per second
Sun	17 years	5 months	8 minutes
Earth's moon	15 days	9 hours	1 second
Mercury	9 years	3 months	4 minutes
Venus	5 years	1 month	2 minutes
Mars	7 years	2 months	3 minutes
Jupiter	70 years	2 years	33 minutes
Saturn	141 years	3 years	1 hour
Uranus	305 years	7 years	2 hours
Neptune	509 years	12 years	4 hours
Pluto†	506 years (890 years)	12 years (213 years)	4 hours (7 hours)
Proxima Centauri (closest star to Earth other than the sun)	—	—	4 years
Sirius	—	—	9 years
Betelgeuse	—	—	310 years
Orion nebula	—	—	1,500 years
Center of the Milky Way galaxy (the galaxy to which Earth belongs)	—	—	27,000 years
Andromeda galaxy (closest external galaxy)	—	—	2 million years

*The estimates given are very rough. For ease of calculation, they assume that the celestial bodies are not in motion (which is of course not the case) and that each body remains at about its minimum possible distance from Earth. Also, in reality a jet could not escape from Earth's gravity and spacecraft have more complex trajectories.

†Pluto's orbit crosses Neptune's, so that Pluto's minimum possible distance from Earth is actually a bit less than Neptune's. However, its mean and maximum possible distances from Earth are much greater than Neptune's. Travel time estimates to Pluto at its maximum distance from Earth are provided in parentheses, since Pluto's distance from Earth varies so much.

cloud objects might be altered is by the gravity of passing stars.

The Kuiper belt and the Oort cloud, then, can be considered vast reservoirs of comets. The Kuiper belt is a doughnut-shaped region of several millions of comet nuclei. They orbit the sun from beyond the orbit of Neptune, mainly between about 30 and 50 AU from the sun. The region for the most part lies fairly close to the plane of the sun's equator (the plane in which the planets orbit). Some Kuiper belt objects have almost circular orbits that fall near this plane.

A composite of 16 exposures tracks the path of the Kuiper belt object called Quaoar. Its small size and great distance from Earth make it very hard to photograph. The icy small body is about half the size of Pluto. It orbits the sun from about 4 billion miles (6.5 billion kilometers) away.

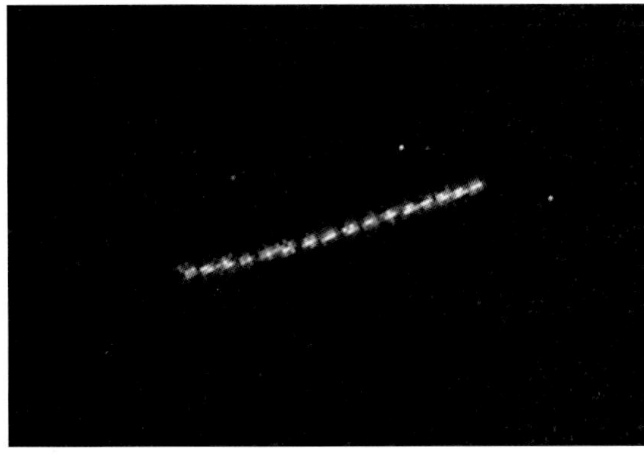

NASA

Others have very elongated, highly inclined orbits. Among the latter group are Eris and Pluto, the largest known members of the Kuiper belt.

The Kuiper belt is thought to be the source of most of the short-period comets, or those that complete an orbit around the sun in less than 200 years, and especially those that take less than 20 years to circle the sun once. The comet nuclei called Centaur objects are also believed to have originated in the Kuiper belt. This group of icy objects is found mainly between the orbits of the outer planets, or about 5 to 30 AU from the sun.

Most of the long-period comets, or those that take more than 200 years to orbit the sun, are thought to come from the Oort cloud. Most of these comets have very long orbital periods, with some taking many millions of years to circle the sun once. The Oort cloud lies much farther from the sun than the Kuiper belt does, extending from perhaps 20,000 to 100,000 AU from the sun. It is not a flat ring but a roughly spherical shell of probably many billions of icy small bodies orbiting in all directions.

In the middle of the 20th century, astronomers first postulated that the Kuiper belt and the Oort cloud must exist. However, because of their great distance from Earth, Kuiper belt objects were not directly detected until the 1990s, when sensitive enough light detectors and telescopes became available. The first object in the belt to be detected was discovered in 1992, and many more large objects have since been found. (Astronomers did not begin to consider Pluto a Kuiper belt object until several other similar objects were found in the belt.) However, none of the much more distant small bodies in the Oort cloud has been seen directly. (*See also* Comet; Astronomy.)

Interplanetary Medium

In the spaces between the planets and other bodies lie vast stretches of extremely thinly distributed matter called the interplanetary medium. The matter includes tiny particles called interplanetary dust or micrometeorites, along with electrically charged particles, tiny amounts of hydrogen gas, and cosmic rays.

The relatively small amounts of interplanetary dust appear to be orbiting the sun in a disk that extends throughout the solar system in or near the plane of the planets' orbit. On a clear night a faint glow is visible in the sky along the line of the zodiac, following the setting sun or preceding the rising sun. This glow can be almost as bright as the Milky Way. It is caused by sunlight reflected by the interplanetary dust. Astronomers estimate that about 30,000 tons of the dust enter Earth's upper atmosphere each year. Spacecraft have detected such dust particles in space nearly as far as the orbit of Uranus. Most of the interplanetary dust is thought to come from the collisions of asteroids and from comets, which lose matter when they approach the sun.

The sun itself contributes much material to the vast spaces between the planets and other bodies. Along with the radiation that continuously leaves its surface, the sun gives off a stream of electrically charged particles—mostly electrons and protons (plasma). This flow of particles is the solar wind. The solar wind spreads beyond the planets to the heliopause, which is the boundary between the interplanetary medium and the interstellar medium, the diffuse matter between the stars. The part of the solar wind that encounters Earth causes the auroras, or the northern and southern lights. The solar wind causes auroras on other planets, too.

The interplanetary medium also includes cosmic rays, which are high-speed, high-energy particles (nuclei and electrons). Some of the cosmic rays come from the sun, but most originate outside the solar system. (*See also* Sun; Astronomy.)

FORMATION AND FUTURE OF THE SOLAR SYSTEM

Astronomers believe that the solar system formed as a by-product of the formation of the sun itself some 4.6 billion years ago. According to the prevailing theory, the sun and its many satellites condensed out of the solar nebula, a huge interstellar cloud of gas and dust. The solar system began forming when the gravity of this interstellar cloud caused the cloud to start contracting and slowly spinning. This could have been caused by random fluctuations in the density of the cloud or by an external disturbance, such as the shock wave from an exploding star.

As the interstellar cloud squeezed inward, more and more matter became packed into the center, which became the protosun (the material that later developed into the sun). The contraction caused the cloud to spin faster and faster and to flatten into a disk. Eventually, the center of the cloud collapsed so much that it became dense enough and hot enough for nuclear reactions to begin, and the sun was born.

Meanwhile, away from the center the gas and dust in the spinning disk cooled. Solid grains of silicates and other minerals, the basis of rocks, condensed out of the gaseous material in the disk. Farther from the center, where temperatures were lower, ices of water, methane, ammonia, and other gases began to form. The spinning material in the disk

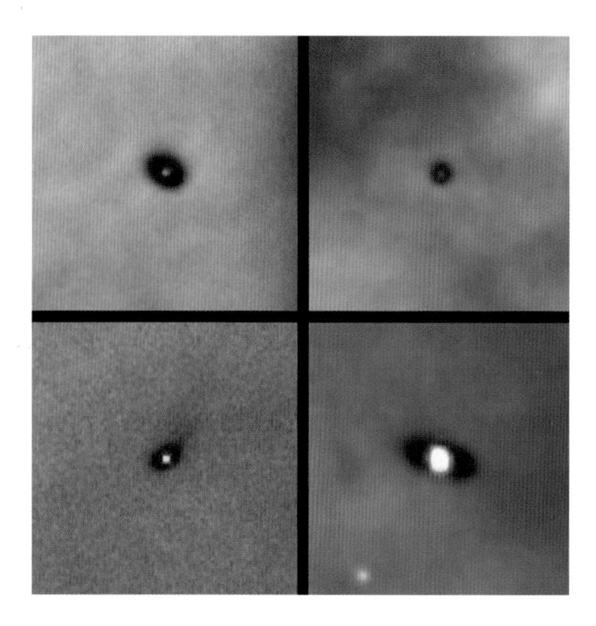

Disks of gas and dust around young stars in the Orion nebula some 1,500 light-years away may be new solar systems in the making. The planets of our solar system are thought to have arisen from such a protoplanetary disk about 4.6 billion years ago. Four images (left) taken by the Hubble Space Telescope show disks surrounding four different stars that are only about a million years old. The central glow in each image is the star. The disks range in size from about two to eight times the diameter of our solar system. Two Hubble images (below) show two views of another newly formed star in Orion. Its protoplanetary disk is seen edge-on, so the star is blocked from direct view. This disk is some 17 times the diameter of our solar system.

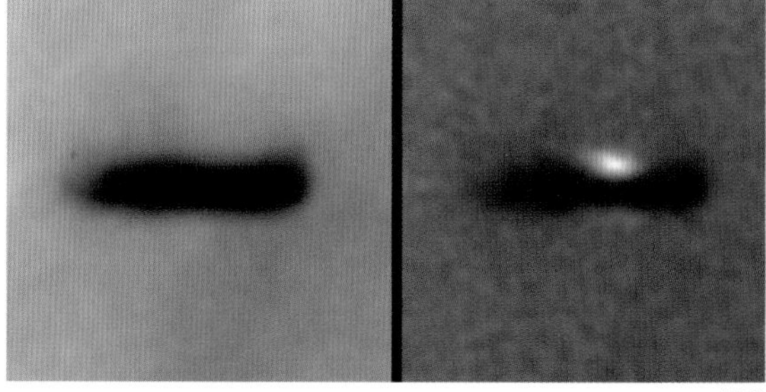

NASA

Properties of Major Objects of the Solar System

Object	Distance from Sun (AU)*	Mean Density (grams/centimeter³)†	Mass (Earth masses)‡	General Composition
Sun	—	1.4	330,000	hydrogen, helium
Mercury	0.4	5.4	0.055	iron, silicates
Venus	0.7	5.2	0.82	silicates, iron
Earth	1	5.5	1	silicates, iron
Earth's moon	1	3.3	0.012	silicates
Mars	1.5	3.9	0.11	silicates, iron, sulfur
Asteroids	2–5 (main belt and Trojans)	typically 1–4	total less than 0.001	silicates, iron
Ceres	2.8	2.2	0.00015	silicates, water ice
Jupiter	5.2	1.3	320	hydrogen, helium
Io (moon of Jupiter)	5.2	3.6	0.015	silicates, sulfur
Europa (moon of Jupiter)	5.2	3.0	0.008	water ice, silicates
Ganymede (moon of Jupiter)	5.2	1.9	0.025	water ice, silicates
Callisto (moon of Jupiter)	5.2	1.8	0.018	water ice, silicates
Saturn	9.5	0.7	95	hydrogen, helium
Titan (moon of Saturn)	9.5	1.9	0.02	water ice, silicates, organics
Centaur objects	5–30 (mainly between orbits of Jupiter and Neptune)	unknown	total possibly 0.0001	presumed similar to that of comets: water ice, other ices, silicates
Uranus	19.2	1.3	14.5	ices, silicates, hydrogen, helium
Neptune	30.1	1.6	17	ices, silicates, hydrogen, helium
Triton (moon of Neptune)	30.1	2.0	0.0036	water ice, silicates, organics
Kuiper belt objects	30–50 (main concentration)	typically 0.3–2	total possibly as much as 0.5	presumed similar to that of comets: water ice, other ices, silicates
Pluto	39.5	2.0	0.0025	water ice, silicates, organics
Eris	67.7	—	—	perhaps similar to Pluto: water ice, silicates, organics
Oort cloud objects	20,000–100,000	typically 0.3–2	total possibly 10–300	presumed similar to that of comets: water ice, other ices, silicates, organics

*Average distance, except where ranges are given. One astronomical unit (AU) is the average distance between Earth and the sun, about 93 million miles (150 million kilometers).
†By comparison, the density of water is 1 gram/centimeter³.
‡Earth's mass is about 6.58×10^{21} tons (5.98×10^{24} kilograms).

collided and began to stick together, forming larger and larger objects. Ultimately, some of the clumped-together objects grew huge and developed into planets. The inner planets formed mostly from chunks of silicate rock and metal, while the outer planets developed mainly from ices. Smaller chunks of matter and debris that did not get incorporated into the planets became asteroids, in the inner part of the solar nebula, and comet nuclei, in the outer part of the nebula. At some point after matter in the nebula had condensed and clumped into objects, the intensity of the solar wind seems to have suddenly increased. This blew much of the rest of the gas and dust off into space.

This general formation process is thought not to be unique to the solar system but rather how stars and planets throughout the universe develop. Astronomers have detected disks of matter surrounding newly formed stars.

The future of the solar system probably depends on the behavior of the sun. If current theories of stellar evolution are correct, the sun will have much the same size and temperature for about 5 billion more years. By then, all of the hydrogen in its core will have been used up. Other nuclear reactions will begin in a shell around the core. Then the sun will grow much brighter and larger, turning into a red giant and expanding beyond the orbit of Venus, perhaps even engulfing Earth. Much later, when all its nuclear energy sources are exhausted, the sun will cool down, evolving into a white dwarf star. Around it will orbit the remaining planets. They will have turned into frozen chunks, orbiting their shrunken star. (*See also* Astronomy.)

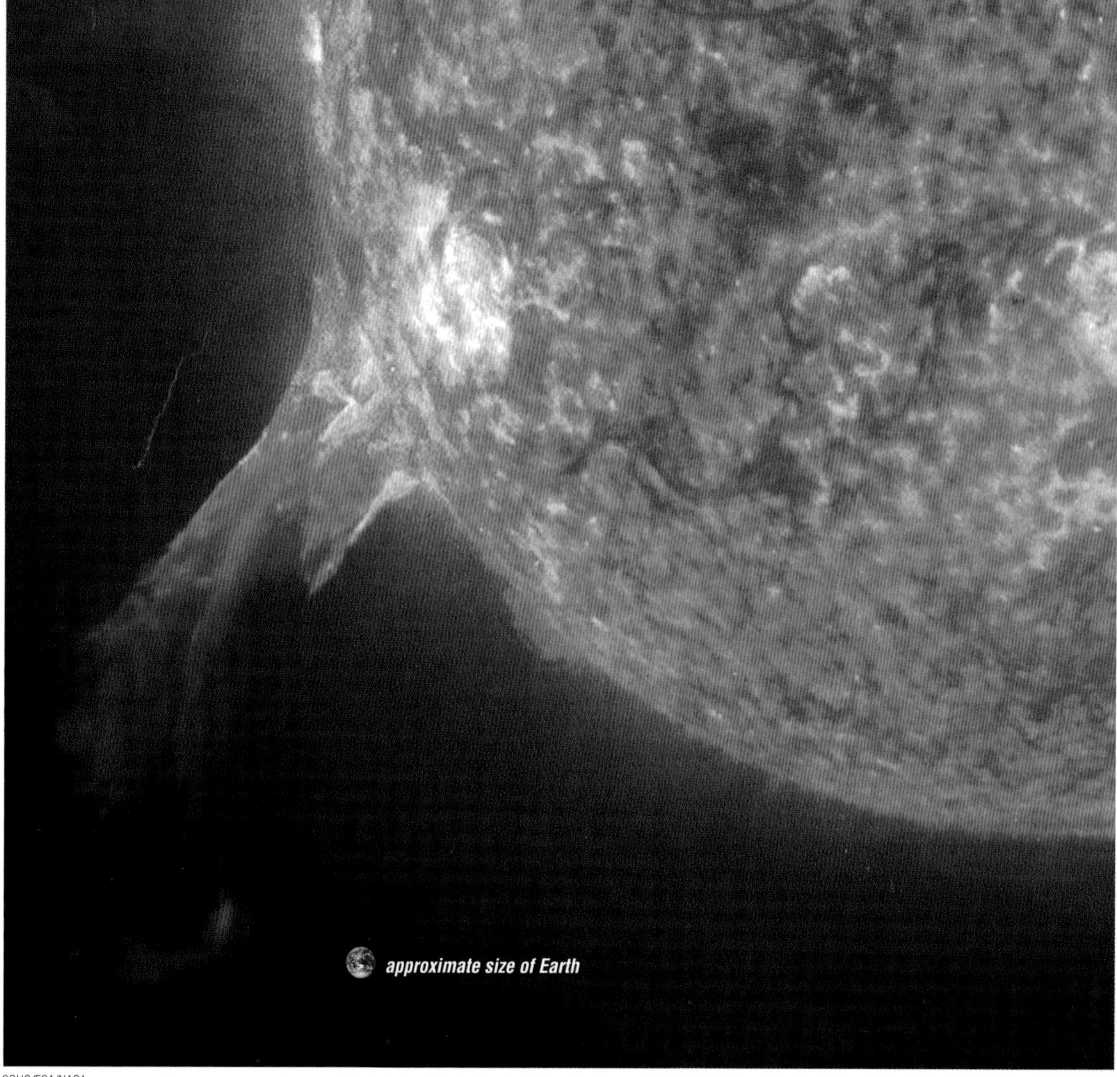

SOHO/ESA/NASA

A huge cloud of gas called a prominence erupts from the sun. An image of Earth has been superimposed to show how enormous the sun is in comparison. Hotter areas on the sun appear in bright white, while cooler areas are red. The image is in false color and was taken in extreme ultraviolet light by the Solar and Heliospheric Observatory (SOHO) orbiting spacecraft.

SUN

Although the sun is a rather ordinary star, it is very important to the inhabitants of Earth. The sun is the source of virtually all Earth's energy. It provides the heat and light that make life on Earth possible. Yet Earth receives only about half a billionth of the energy that leaves the sun. The sun is a huge ball of hot gases. Like other stars, it produces enormous amounts of energy by converting hydrogen to helium deep within its interior.

Because this energy is so intense, there are some dangers in staring at the sun. Radiation from the sun's rays can damage one's eyes, so one should never look directly at the sun with unaided eyes or with a telescope (unless it is equipped with a special solar filter). Dark glasses and smoked glass provide no protection. One safe way to observe the sun is to project its image through a pinhole or telescope onto a white screen or white cardboard.

Position in the Solar System

The sun lies at the center of the solar system. It contains more than 99 percent of the system's mass. The immense pull of its gravity holds the planets, dwarf planets, asteroids, comets, and other bodies in orbit around it. The average distance between the sun and Earth is roughly 93 million miles (150 million kilometers). Sunlight travels through space at about 186,282 miles (299,792 kilometers) per second, so a ray of sunlight takes only about 8 minutes to reach Earth. However, light from the next nearest star, Proxima Centauri, takes more than four years to arrive. The sun is in the outer part of the Milky Way galaxy. Light from the center of the galaxy takes many thousands of years to reach Earth. Because the sun is so close to Earth, it seems much larger and brighter than other stars. It is the only star whose surface details can be observed from Earth.

Basic Properties

Stars vary greatly in size and color. They range from giant stars, which are much larger than the sun, to dwarf stars, which can be much smaller than the sun. In color they range from whitish blue stars with very high surface temperatures (more than 30,000 Kelvin, or 53,500° F) to relatively cool red stars (less than 3,500 K, or 5,840° F). (The Kelvin temperature scale uses degrees of the same size as Celsius degrees but is numbered from absolute zero, –273.15° C.) The sun is a yellow dwarf star, a kind that is common in the Milky Way galaxy. Its surface temperature is about 5,800 K (10,000° F). Its diameter is about 864,950 miles (1,392,000 kilometers), which is about 109 times the diameter of Earth. Its volume is about 1,300,000 times as great as Earth's, and its mass is some 333,000 times Earth's.

More than 90 percent of the sun's atoms are hydrogen. Most of the rest are helium, with much smaller amounts of heavier elements such as carbon, nitrogen, oxygen, magnesium, silicon, and iron. By mass, the sun is about 71 percent hydrogen and 28 percent helium.

The sun has no fixed surface. It is much too hot for matter to exist there as a solid or liquid. Rather, the sun's matter consists of gas and plasma, a state in which gases are heated so much that the electrons are stripped away from their atomic nuclei. The heated gas is said to be ionized because it consists of a group of ions, or electrically charged particles. The free electrons carry a negative charge, and the atomic nuclei carry a positive charge.

Like the planets, the sun rotates. Because the sun is not solid, different parts of it rotate at different rates. The parts of the surface near the equator spin the fastest, completing one rotation about every 25 Earth days, compared to 36 days for parts near the surface.

Structure and Energy Production

The sun can be divided into several different layers. Energy is produced in the dense, hot, central region, which is called the core, and travels outward through the rest of the interior. The surface, or the part of the sun that is visible from Earth in ordinary light, is called the photosphere. It emits most of the light and heat that reach Earth. The surface is the innermost part of the solar atmosphere. The atmosphere also has a thin middle layer, called the chromosphere, and a large outer layer, the corona. The corona gives rise to a flow of charged particles called the solar wind that stretches beyond Earth and the other planets.

Energy and balance of forces. The sun looks like a burning sphere. It is too hot, however, for an Earth-type chemical reaction such as burning to occur there. Besides, if burning produced its energy, it would have run out of fuel very long ago.

Various theories have been advanced to explain the sun's tremendous energy output. All the bits of matter in the sun exert gravitational attraction on each other. One 19th-century theory said that this gravitational attraction causes the sun to shrink and its matter to become more tightly packed. This process, called gravitational contraction, could release a great deal of energy. However, gravitational contraction would produce energy for only 50 million years at most, while the sun's age must be at least as great as Earth's age of about 4.6 billion years.

In the 20th century atomic theory finally provided an explanation. Scientists now agree that thermonuclear reactions are the source of solar energy. Albert Einstein's theoretical calculations showed that a small amount of mass can be converted to a great amount of energy. Reactions in the sun's core convert almost 5 million tons of matter into enormous amounts of energy—3.86×10^{33} ergs—every second. The vast amount of matter in the sun can provide the "fuel" for billions of years of atomic reactions. Astronomers believe that the sun is nearly halfway through its "lifetime" of 10 billion years.

The sun's thermonuclear reactions also keep the star from squeezing inward. While the sun's gravity exerts a huge inward pull, the energy it produces exerts a huge outward pressure. At this stage in the sun's life, these forces balance each other out, so that the sun neither collapses under its own weight nor expands.

Core. The sun's core is an extremely hot, dense mass of atomic nuclei and electrons. Its temperature is about 15,000,000 K (27,000,000° F), and it is thought to be some 150 times as dense as water. The pressure is enormous. Normally, protons in atomic nuclei repel each other because they have the same electrical charge. Under the great density and pressure in the sun's core, however, nuclei can collide and fuse into new and heavier nuclei. This is a type of thermonuclear reaction called a fusion reaction.

Facts About the Sun

Diameter at Equator. 864,950 miles (1,392,000 kilometers), or about 109 times Earth's diameter.

Mass. 2 octillion tons (2×10^{33} grams), or about 333,000 times Earth's mass.

Volume. 50 octillion cubic feet (1.4×10^{33} cubic centimeters), or about 1,300,000 times Earth's volume.

Average Density. 1.4 grams per cubic centimeter, or about 0.255 Earth's average density.

Rotation Period. 25–36 Earth days, depending on latitude.

Temperature at Core. 15,000,000 K (27,000,000° F).*

Temperature at "Surface" (Photosphere). 5,800 K (10,000° F).

*In the Kelvin scale the zero point is absolute zero, which is –273.15° C. To convert temperatures from degrees Kelvin (K) to degrees Celsius (C), subtract 273.15.

The basic fusion process in the sun involves a series of reactions in which four hydrogen nuclei are ultimately converted into one helium nucleus. The mass of the helium nucleus is about 0.7 percent less than that of the four hydrogen nuclei. This 0.7 percent of the mass is changed into energy. Every second the sun converts almost 700 million tons of hydrogen into about 695 million tons of helium. Nearly 5 million tons of mass—0.7 percent of 700 million tons—are converted to energy. Some of this energy heats the plasma in the core and some escapes into space as nearly massless, electrically neutral particles called neutrinos. Some of the energy is in the form of gamma-ray photons. These photons travel outward from the core through a zone in which the energy is carried mainly by radiation. Ultimately, the energy is emitted at the surface in many different wavelengths.

Radiation and convection zones. The radiation (or radiative) zone is very dense and opaque. Photons take a long, randomly zigzagged path through it. It takes the energy hundreds of thousands of years, or by some estimates millions of years, to pass through this zone. Gamma-ray photons can travel only a tiny distance before colliding with other particles and being scattered. Farther outward, the photons collide with atoms, which absorb the energy and then reradiate it. The atoms reradiate the energy at progressively longer wavelengths and lower energies. By the time the energy leaves the sun, much of it is in the form of visible light and infrared radiation (heat).

Surrounding the radiation zone is a cooler region called the convection zone. It takes up about the outer 30 percent of the sun's interior. In this zone great currents of hot gases bubble upward, while cooler, denser matter sinks (like the circulation in a pot of oatmeal or water boiling on the stove). These currents, called convection currents, transport energy to the surface of the sun, the photosphere.

Photosphere. The photosphere (meaning "sphere of light") is the lowest layer of the sun visible from Earth. This thin layer is the lowest level in the sun's atmosphere. Energy finally escapes the sun from the photosphere, so it is significantly cooler than the solar interior. The temperature at the visible surface is about 5,800 K (10,000° F). The solar atmosphere is also dramatically less dense than the interior.

The photosphere has a definite texture. It is covered with granules, or luminous grainlike areas separated by dark areas. Granules are continually forming and disappearing. Their grainlike structure results from the convection currents that bring hot gases up to the photosphere. Each granule is a convection cell that measures several hundred miles across. The hot upwelling matter appears bright, while the cooler sinking matter appears dark. Periodically, larger darker blotches called sunspots (discussed below) appear on the photosphere.

The entire surface of the sun continually oscillates, or pulses, up and down. These vibrations result from the motions of millions of sound waves and other kinds of waves that are trapped inside the sun. The waves travel inward and outward through the sun's interior and cause the surface to vibrate, like the surface of a ringing bell.

Chromosphere. The layer of the atmosphere above the photosphere is called the chromosphere (meaning "sphere of color"). It is visible as a thin reddish ring around the edge of the sun during total solar eclipses, when the much brighter photosphere is blocked from view. It can also be observed with telescopes with a certain filter (hydrogen alpha). The chromosphere is hotter than the photosphere, and its temperature generally rises with altitude. It is marked by countless jets or small spikes of matter called spicules that continually form and disappear, rising up and falling back down within minutes. Much of the sun's "weather" also takes place in the chromosphere. This includes the violent eruptions called solar flares (discussed below).

Corona. The chromosphere is surrounded by a faintly luminous, extremely thin outer atmosphere called the corona (meaning "crown"). As the corona is a million times dimmer than the sun's disk, it is usually invisible. It can be seen only when the light of the photosphere is blocked, as in a total solar eclipse or with a special type of telescope. The corona then appears as a silvery halo with long arcs and streamers.

Much or all of the corona's volume consists of loops and arcs of hot plasma. Counterintuitively, the corona is much hotter than the surface of the sun. Solar scientists think that energy from the solar magnetic fields heats the corona, but the mechanism for this is not completely clear. The corona is typically about 2,000,000 K (3,600,000° F) at its inner levels. However, this temperature is a measure of the energy of the individual particles in the plasma. The corona's density is so low—the particles of matter are spread out so widely—that the corona does not actually

Hot jets of matter called spicules rise in the sun's chromosphere. The spicules appear as short, dark features throughout the image, especially at right. Sunspots are visible at upper left. The image was taken with a hydrogen-alpha filter on the Royal Swedish Academy of Sciences' Swedish 1-meter Solar Telescope (SST) on the Spanish island of La Palma. The color was added during processing.

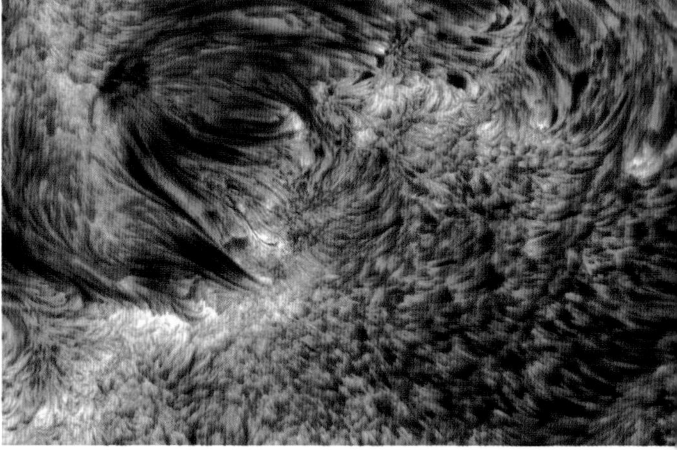

LOCKHEED MARTIN/Solar and Astrophysics Lab

During a total solar eclipse, the sun's corona appears as a white, glowing ring around the black disk of the moon, which obscures the rest of the sun. The corona is the outer layer of the sun's atmosphere. It is not visible unless the bright disk of the sun is blocked out.

produce much heat. A meteor traveling through the corona does not burn up, as commonly happens in Earth's much cooler but much denser atmosphere.

Although the corona is relatively faint in visible light, it strongly emits radiation at extreme ultraviolet and X-ray wavelengths. However, areas of the corona periodically appear dark at those wavelengths. The corona is extremely thin in these dark areas, called coronal holes. The magnetic fields of the coronal holes open freely into space, and charged particles stream out along the magnetic lines of force. (Magnetic lines of force, or magnetic field lines, show the direction and strength of a magnetic field. Charged particles can easily move through space along these lines but not across them.)

Solar wind. Spacecraft in interplanetary space have encountered streams of highly energetic charged particles originating from the sun. These streams, called the solar wind, flow radially outward from the sun's corona through the solar system and extend beyond the orbits of the planets. These particles are continuously released, but their numbers increase greatly following solar flares and other eruptions.

The solar wind is a plasma consisting chiefly of a mixture of protons and electrons plus the nuclei of some heavier elements in smaller numbers. The solar wind may be formed when the hot coronal plasma expands. The particles are accelerated by the corona's high temperatures to speeds great enough to allow them to escape from the sun's gravitational field.

The fastest streams of the solar wind come from particles that flow from the coronal holes. These particles travel as fast as about 500 miles (800 kilometers) per second. Other streams of the solar wind reach speeds as high as about 250 miles (400 kilometers) per second. These streams usually originate in regions near the solar equator.

As they flow outward, the particles of the solar wind carry part of the sun's magnetic field along with them. Because of the sun's rotation and the steady outflow of particles, the lines of the magnetic field carried by the solar wind trace curves in space. The solar wind is responsible for deflecting the tails of passing comets away from the sun. Luckily for Earth, the planet's magnetic field shields it from the radiation of the solar wind. When the streams of particles encounter Earth's magnetic field, a shock wave results.

Magnetic Fields and Solar Activity

The sun's magnetic activity is quite complex. Rapid, large fluctuations occur in numerous strong local magnetic fields that are threaded through the sun's atmosphere. Magnetic activity shapes the atmosphere and causes disturbances there called solar activity. This activity includes sunspots and violent eruptions. Overall, solar activity follows about an 11-year cycle, in which the numbers of sunspots and other disturbances increase to a maximum and then decrease again. The sun seems to have a weak global magnetic field. Once each 11-year cycle, the north and south poles of the field switch polarity.

Sunspots. Periodically, darker, cooler blotches called sunspots temporarily appear on the sun's surface. Sunspots are areas where very strong local magnetic fields interfere with the normal convection activity that brings heat to the surface. The spots usually appear in pairs or groups of pairs. Each spot typically has a darker, circular center, called the umbra, surrounded by a lighter area, the penumbra. The umbras are about 2,000 K (3,100° F) cooler than the photosphere around them (which means that they are still very hot). Sunspots vary greatly in size but are always small compared to the size of the sun. When they appear in groups, they may extend over tens of thousands of miles. They last from tens of minutes to a few days or even months.

Regular observations of sunspots have been made since 1750. They reveal that the spots appear and disappear in a cycle, and that they are limited to the two zones of the sun contained between about latitudes 40° and 5° of its northern and southern

A group of sunspots appear against granules on the surface of the sun in a very sharp image taken with the Swedish 1-meter Solar Telescope (SST) on the Spanish island of La Palma. Each sunspot has a dark central umbra surrounded by a lighter penumbra. The color was added during processing. The observations were made by Göran Scharmer, and the image was processed by Mats Löfdahl.

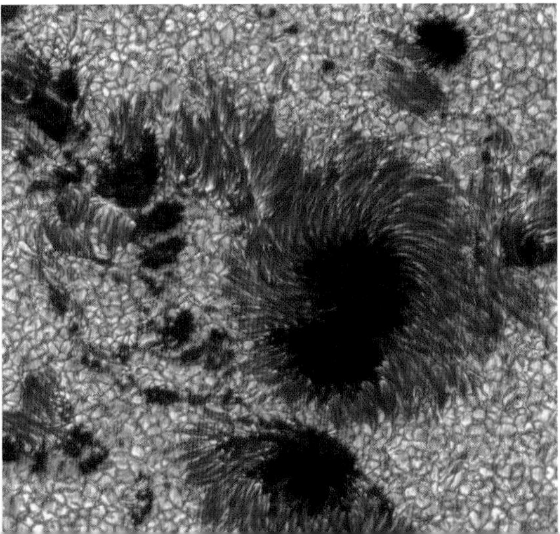

hemispheres. As mentioned above, the cycle lasts an average of about 11 years. At the beginning of a cycle a few spots appear at around 35° latitudes. Then they rapidly increase in number, reaching a maximum in the course of around five years. At the same time, the spots get closer and closer to the equator. During the next six years their number decreases while they continue to approach the equator. The cycle then ends, and another cycle starts.

In the early 20th century George E. Hale observed that certain photographs of sunspots showed structures that seemed to follow magnetic lines of force. Often a pair of sunspots appeared to form the north and south poles of a magnetic field. Hale was finally able to establish that sunspots are indeed seats of magnetic fields. In addition, from one 11-year cycle to the next, a total reversal of the sunspots' polarity occurs in the two solar hemispheres. In other words, the north pole of a magnetic field associated with a sunspot pair becomes the south pole, and vice versa. A magnetic cycle of sunspots lasts an average of 22 years, since it encompasses two 11-year cycles.

Flares, prominences, and coronal mass ejections. A more violent phenomenon is the solar flare, a sudden eruption in the chromosphere above or near sunspot regions. Flares release magnetic energy that builds up along the boundaries between negative and positive magnetic fields that become twisted. The flares usually form very rapidly, reaching their maximum brilliance within minutes and then slowly dying out. They emit huge amounts of radiation at many different wavelengths, including X-rays and gamma rays, as well as highly energetic charged particles.

Features called prominences also form along sharp transitions between positive and negative magnetic fields. Early astronomers noticed huge red loops and streamers around the black disk of solar eclipses. These prominences are areas of relatively cooler, denser plasma suspended like clouds through the hot, low density corona. Magnetic lines of force hold the plasma in place. Prominences appear as bright regions when seen extending from the solar disk but as long, dark threadlike areas when seen against the disk. The dark areas are also called filaments.

An image taken in extreme ultraviolet light reveals a solar prominence lifting off the sun. A prominence is a region of cooler, denser plasma suspended in the corona by the sun's magnetic field lines. The false-color image was captured by the Transition Region and Coronal Explorer (TRACE) orbiting satellite.

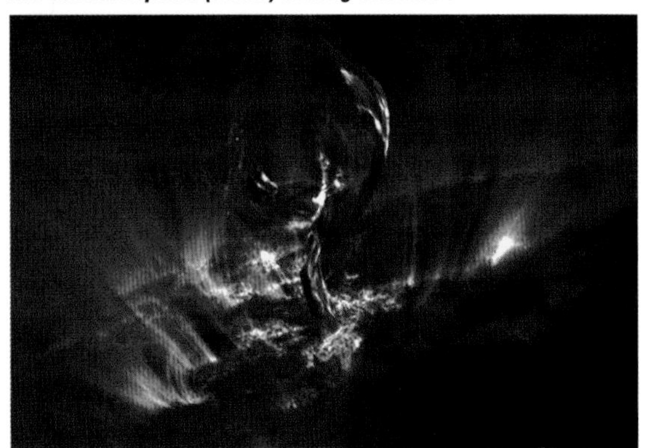

TRACE/NASA

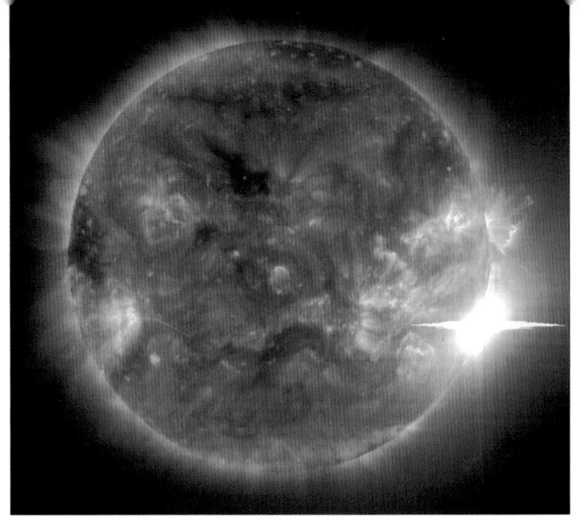

SOHO/ESA/NASA

One of the strongest solar flares ever detected appears at right in an extreme ultraviolet (false-color) image taken by the Solar and Heliospheric Observatory (SOHO) orbiting spacecraft. Such powerful flares, called X-class flares, release intense radiation that can temporarily cause blackouts in radio communications all over Earth. The flare occurred on Nov. 4, 2003.

Long-lived, or quiescent, prominences may keep their shape for months. They form at the boundaries between large-scale magnetic fields. Prominences in active regions associated with sunspots are short-lived, lasting only several minutes to a few hours. When prominences become unstable, they may erupt upward. These eruptions are significantly cooler and less violent than solar flares.

A type of violent eruption called coronal mass ejections also occurs in the corona. The corona sometimes releases enormous clouds of hot plasma into space. Like solar flares, these coronal mass ejections release energy built up in solar magnetic fields. They usually last hours, however, while the rapid eruptions from flares typically last only minutes. Like other kinds of solar activity, coronal mass ejections are most common during the solar maximum. Scientists believe that flares, prominence eruptions, and coronal mass ejections are related phenomena. Their relationship is complex, however, and not yet fully understood.

In any case, the sun's violent eruptions have concrete effects on Earth. Large solar flares and coronal mass ejections shower Earth with streams of high-energy particles that can cause geomagnetic storms. These storms can disrupt communications satellites and radio transmissions and cause surges in power transmission lines. They also create auroras (the northern and southern lights) near the poles.

Studying the Sun

When a ray of sunlight, which appears white, is passed through a prism or a diffraction grating, it spreads out into a series of colors called a spectrum. Scientists analyze this spectrum to determine what chemicals make up the sun as well as their abundance, location, and physical states.

Two views of the sun taken at about the same time show active regions in the atmosphere (left) and corresponding sunspots on the surface below (right). The false-color image at left was taken in extreme ultraviolet light, while the image at right was taken in white (visible) light. Both were captured by the Solar and Heliospheric Observatory (SOHO) orbiting spacecraft.
SOHO/ESA/NASA

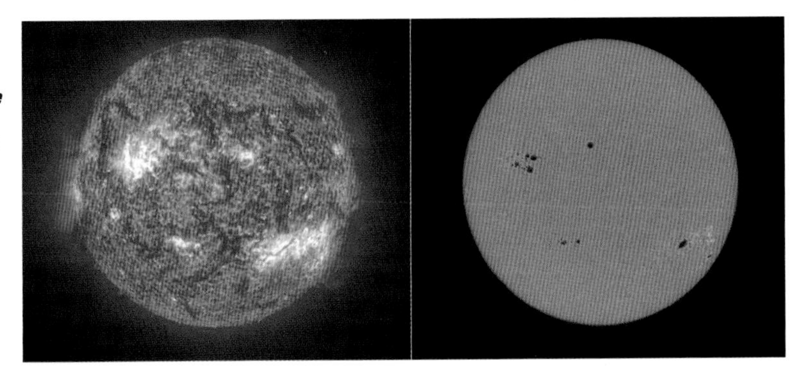

In 1814 Joseph von Fraunhofer began a thorough study of the sun's spectrum. He found that it was crossed by many dark lines, which are now called absorption lines or Fraunhofer lines. Meanwhile, other scientists had been studying the light emitted and absorbed by elements in the gaseous state when they were heated in the laboratory. They discovered that each element always produced a set of bright emission lines associated with that element alone.

Scientists now believe that the dark lines in the sun's spectrum represent elements in the sun's atmosphere. The line that Fraunhofer had called D, for example, was shown to have the same position in the spectrum as the brilliant line that sodium gave off when it was heated in the laboratory. The lines are dark because the elements in the sun's atmosphere absorb the bright lines given off by the same element on the sun's disk. Studying the lines of the sun's spectrum, then, provides a way to study the composition of the sun's atmosphere. Almost all elements known on Earth have been shown to exist on the sun.

The telescope has been used in solar studies since 1610. The tower solar telescope was later invented for use in solar studies. Its long focal length can give very large images of the sun. The coronagraph, another special telescope, is used to examine the sun's atmosphere. The instrument blocks the direct light from the sun's disk, allowing the much dimmer corona to be viewed.

Different layers and features of the sun's atmosphere emit radiation more strongly at different wavelengths. Solar scientists use instruments that detect radiation at specific wavelengths to study the different parts. For example, the spectroheliograph and the birefringent filter can limit the light that passes through them to a very small range of wavelengths, such as the red light emitted by hydrogen (known as hydrogen alpha) or the violet light of calcium. The photosphere can be seen well in visible light. The chromosphere and the prominences in the corona emit much of their radiation in hydrogen alpha. X-rays and extreme ultraviolet are often used to study the corona and solar flares, which emit much radiation at those wavelengths.

On the ground, the effectiveness of a telescope is limited because the Earth's atmosphere absorbs much of the sun's radiation at certain wavelengths. Many orbiting solar observatories and other spacecraft have been launched above the terrestrial atmosphere to study the sun. Special instruments aboard the craft are used to photograph and measure a variety of solar properties and features, including the sun's magnetic field and the charged particles of the solar wind.

Since the sun's interior cannot be directly seen, solar scientists infer its properties from the behavior of the atmosphere. One method involves studying the oscillations of solar waves. As mentioned above, trapped sound waves moving inside the sun constantly cause some parts of the sun to move outward and some parts to move inward. In a field called helioseismology, scientists detect and analyze the properties of these waves, including the patterns of their oscillations, or modes. In general, the waves travel in one direction until abrupt changes in density or temperature within the sun's layers bend them back in the opposite direction. Helioseismologists use this information to develop models of the structure and motions of the solar interior. This is similar to the way geologists study the seismic waves produced by earthquakes in order to map Earth's interior.

The sun violently ejects a bubble of hot plasma in a very large coronal mass ejection (CME), at upper right. The material erupts into space at about 1–2 million miles (1.5–3 million kilometers) per hour. The image was taken with a coronagraph, an instrument that blocks the solar disk to reveal the much dimmer corona. The red disk in the center is part of the instrument; the white circle indicates the size and position of the sun's disk. The false-color image was taken by the Solar and Heliospheric Observatory (SOHO) orbiting spacecraft.
SOHO/ESA/NASA

The solar system's four inner planets are much smaller than its four outer planets, and all eight are dwarfed by the sun they orbit. The sizes of the bodies are shown to scale, though the distances between them are not.

PLANET

The relatively large natural bodies that revolve in orbits around the sun or other stars are called planets. The term does not include small bodies such as comets, meteoroids, and asteroids, many of which are little more than pieces of ice or rock.

The planets that orbit the sun are part of the solar system, which includes the sun and all the bodies that circle it. The sun governs the planets' orbital motions by gravitational attraction and provides the planets with light and heat. Ideas about what makes a planet a planet have changed over time. According to the current definition, there are eight planets in the solar system. In order of increasing mean distance from the sun, the planets are Mercury, Venus, Earth, Mars, Jupiter, Saturn, Uranus, and Neptune. Pluto was considered the solar system's ninth and outermost planet from 1930 until 2006, when it was reclassified as a dwarf planet.

The concept of what constitutes a planet changed in another way in the late 20th century when astronomers began discovering distant planets that orbited stars other than the sun. Previously, the only planets that had been known were the planets of the solar system.

What Is a Planet?

Mercury, Venus, Mars, Jupiter, and Saturn, all of which can be seen without a telescope, have been known since ancient times. The ancient Greeks used the term planet, meaning "wanderers," for these five bodies plus the sun and Earth's moon, because the objects appeared to move across the background of the apparently fixed stars. The names of these seven bodies are still connected, in some languages, with the days of the week. In time astronomers learned more about how celestial objects move in the sky, and they recognized that the sun, not Earth, is the center of the solar system. The term planet came to be reserved for large bodies that orbit the sun and not another planet, so the sun and moon were no longer considered planets.

With the aid of telescopes, modern astronomers have discovered hundreds of thousands of additional objects orbiting the sun, including more planets. Although Uranus is sometimes visible with the naked eye, ancient astronomers were unable to distinguish it from the stars. Uranus was first identified as a planet in 1781. The eighth planet, Neptune, was discovered in 1846.

Pluto was initially considered a planet. Indeed, its discovery in 1930 was the result of a major search for a distant ninth planet. Astronomers found that Pluto was unlike the other planets in many ways, including its composition of rock and ice and its tilted and very elongated orbit. For decades it seemed unique. Beginning in the late 20th century, however, advances in telescope technology allowed astronomers to discover many more small icy objects with elongated orbits that lie beyond Neptune. Along with Pluto, these numerous objects orbit the sun in a doughnut-shaped zone called the Kuiper belt. At first it seemed that Pluto was significantly larger than all these objects. But a few Kuiper belt objects were found to be nearly as big as Pluto, and one (which was later named Eris) was slightly larger. This raised the question of whether those bodies should be considered planets if Pluto was.

To address this issue and the growing controversy swirling around it, the group that approves the names of astronomical objects for the scientific community issued an official definition in 2006. The organization, known as the International Astronomical Union (IAU), defined a planet as a celestial body that meets three conditions. First, it must orbit the sun and not be a satellite of another planet. Second, it must be massive enough for its gravity to have pulled it into a nearly spherical shape. Third, it must have "cleared the neighborhood around its orbit," meaning that it must be massive enough for its gravity to have removed most rocky and icy debris from the vicinity of its orbit.

Pluto is not a planet under this definition. It orbits the sun and is nearly round, but it has not cleared away many small Kuiper belt objects from the area around its orbit. The IAU created a new category, called dwarf planet, for those nearly spherical celestial bodies that orbit the sun but that have not cleared the area around their orbits and that are not moons. The first bodies to be designated dwarf planets were Pluto, Eris, and Ceres (which is also the largest asteroid). (*See also* Dwarf Planet; Pluto; Solar System.)

The IAU definitions were themselves controversial. Some scientists welcomed them as appropriate recognition that Pluto is one of the larger objects in the Kuiper belt. Other scientists criticized the definitions' lack of precision and scientific rigor and called for their revision.

Classification of the Planets

The eight planets of the solar system can be divided into two groups—inner planets and outer planets—according to their basic physical characteristics and positions relative to the sun. The four inner planets are Mercury, Venus, Earth, and Mars. They are also called the terrestrial, or Earth-like, planets. These relatively small worlds are composed primarily of rock and metal. With densities ranging from nearly

Photo NASA/JPL/Caltech (NASA photo # PIA00157)

Like the other inner planets of the solar system, Venus has a solid surface. A color-coded radar image reveals its topography. The lowest elevations are colored violet, while the highest elevations appear in red and pink. The large red and pink area at the top is Maxwell Montes, the planet's highest mountain range.

four to five and a half times the density of water, the inner planets have solid surfaces. None of these planets has rings, and only Earth and Mars have moons.

The four outer planets are Jupiter, Saturn, Uranus, and Neptune. They are also called the Jovian, or Jupiter-like, planets. All of them are much bigger than the inner planets, and Jupiter is more massive than the seven other planets combined. Unlike the inner planets, the outer planets have no solid surfaces. Their densities are less than twice the density of water. Jupiter and Saturn are composed mostly of hydrogen and helium in gaseous and liquid form, while Uranus and Neptune consist of melted ices and molten rocks, as well as hydrogen and helium. Each of the four has a massive atmosphere, or surrounding layers of gases,

Properties of the Solar System's Planets, Expressed in Earth Ratios

	Mercury	Venus	Earth	Mars	Jupiter	Saturn	Uranus	Neptune
Average Distance from Sun (AU)	0.4	0.7	1	1.5	5.2	9.5	19.2	30.1
Average Diameter at Equator*	0.38	0.95	1	0.53	11.21*	9.45*	4.01*	3.88*
Mass	0.055	0.82	1	0.11	320	95	14.5	17
Average Density	0.98	0.95	1	0.71	0.24	0.13	0.23	0.30
Average Surface Gravity*	0.38	0.88	1	0.38	2.36*	0.91*	0.89*	1.14*
Average Surface Pressure	~10^{-15}	95	1	0.006	—	—	—	—
Rotation Period (Sidereal)	58.6	243	1	1.03	0.41	0.45	0.72	0.67
Orbital Period (Year on Planet)	0.24	0.62	1	1.88	11.86	29.44	84.01	163.72

*For the outer planets (Jupiter, Saturn, Uranus, and Neptune), which have no solid surfaces, these values are calculated for the altitude at which 1 bar of atmospheric pressure (the pressure of Earth's atmosphere at sea level) is exerted.

(Left) NASA/JPL/University of Colorado; (right) ESO

Saturn's enormous and stunning rings are well known, but all the outer planets have rings. A close-up ultraviolet image (above) taken by the Cassini spacecraft shows the A ring of Saturn. The ringlets shown in turquoise have a greater percentage of water ice than those shown in red. A near-infrared image of Uranus (right) taken from Earth at the Very Large Telescope facility in Chile shows the planet, its rings, and several of its moons. Uranus' rings are nearly impossible to see from Earth in visible light, even with the most powerful telescopes.

of mostly hydrogen and helium. Each of the outer planets also has a ring system and many moons.

Motions of the Planets

The solar system's planets move through space in two basic ways simultaneously. They orbit the sun and rotate about their centers.

Orbit. The eight planets orbit the sun in the same direction as the sun's rotation, which is counterclockwise as viewed from above Earth's North Pole. The planets also orbit in nearly the same plane, so that their paths trace out a large disk around the sun's equator. Mercury's orbit is the most tilted. It is inclined about 7 degrees relative to the ecliptic plane, or the plane in which Earth orbits. The orbital planes of all the other planets are within about 3.5 degrees of the ecliptic plane. By comparison, Pluto's orbit is inclined about 17 degrees.

In the early 1600s German astronomer Johannes Kepler discovered three major laws that govern the motions of planets. The first law describes the shape of their orbits, which are not exactly circular but slightly oval. The orbits form a type of closed curve called an ellipse. Each ellipse has two imaginary fixed points inside the curve called foci, which is the plural of "focus." (The sum of the distance from a point on the curve to one focus and from the same point to the other focus is the same for all points on the ellipse.) The sun lies at one of the two foci of the ellipse of each planet's orbit. Mercury has the most eccentric, or elongated, orbit of the planets, while Venus and Neptune have the most circular orbits. The orbits of Pluto, Eris, and many other Kuiper belt objects and comets are significantly more eccentric.

Kepler's second law describes the velocities of the planets in their orbits. It states that an imaginary line drawn from a planet to the sun sweeps across equal areas in equal periods of time. This means that the planets move faster when their orbits bring them closer to the sun and more slowly when they are farther away. Kepler's third law allows one to calculate a planet's orbital period, or the time it takes the planet to complete one orbit around the sun, if one knows its average distance from the sun, and vice versa. The law states that the square of a planet's orbital period is proportional to the cube of the planet's average distance from the sun.

Rotation. In addition to orbiting the sun, each planet rotates about its axis, an imaginary line that extends through the planet's center and its north and south poles. Most of the planets rotate in the same direction that they orbit. Only Venus and Uranus rotate in the opposite direction, which is called retrograde rotation. The rotational axes of all the planets except Uranus are more or less upright, or perpendicular to the ecliptic plane. Oddly, Uranus' axis is almost parallel to the ecliptic plane, so the planet rotates nearly on its side.

Years, days, and seasons. A planet's motions in space define the length of a year and of a day on that planet. A year is the time it takes the planet to complete one orbit around the sun. Earth's orbital period, for example, is about 365 days, so that is the length of one Earth year.

The length of a planet's day is defined by its rotation in two different ways. A day is the time it takes the planet to rotate on its axis once relative to the sun. Astronomers call this a solar day. During one solar day the sun moves from its noontime position in the sky, sets, and rises to the noon position again. A sidereal day is the time it takes the planet to rotate once relative to distant stars. In most cases, the solar and sidereal days are nearly the same length. Earth's solar day is about four minutes longer than its

Mars's spin axis is tilted about 24.9° relative to the plane in which it orbits. As the planet travels in its orbit, first the northern hemisphere, then the southern hemisphere is tipped toward the sun. As a result, there are four distinct seasons on Mars. The ice caps at the poles alternately grow and shrink as the seasons change.

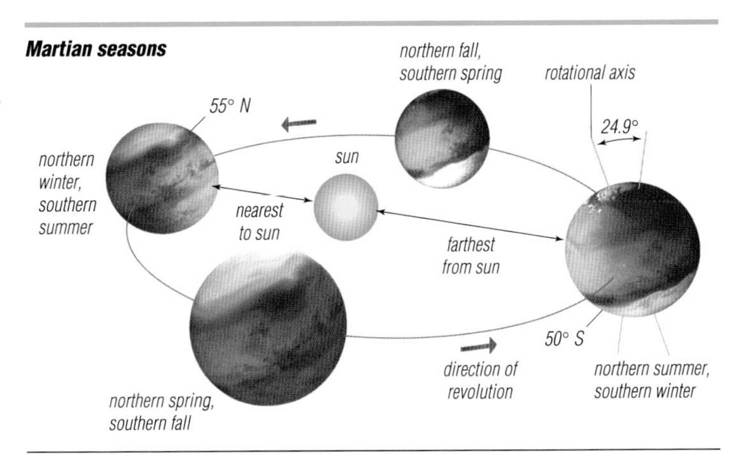

sidereal day, while on Mars and all the outer planets this difference is less than a minute. Mercury and Venus are unusual in that their solar and sidereal days differ by many hours.

Seasons are caused mainly by the tilt of a planet's spin axis relative to its plane of orbit. The variation in the planet's distance from the sun over the course of a year is a secondary factor. The axes of Mercury, Venus, and Jupiter are barely tilted, so they experience little or no seasonal differences in solar illumination and weather.

Earth's axis is tilted about 23.5 degrees. As the planet travels around the sun, first one hemisphere then the other is tipped slightly toward the sun. As a result, the angle of the sunlight reaching the ground at noon varies over the course of the year. The half of Earth that is tipped toward the sun experiences summer, while the half that is tipped away experiences winter. During summer the sun is higher in the sky at noon and the hours of daylight per day are longer, which causes greater heating.

Mars, Saturn, and Neptune have axial tilts that are a bit greater than Earth's—roughly 25 to 30 degrees—so they also have seasons. Because these planets take longer to orbit the sun, the length of each season is much longer than on Earth.

The extreme tilt of Uranus' axis, more than 97 degrees, gives it extreme seasons. For part of the year, the planet's north pole points roughly toward the sun, while the southern hemisphere remains dark day and night. Later in the year conditions are reversed, with the northern hemisphere draped in darkness and the southern hemisphere bathed in light.

Apparent motions. The motions of the planets as observed from Earth, called their apparent motions, are complicated by Earth's own revolution, rotation, and slightly tilted axis. Earth rotates from west to east, so that both the stars and the planets appear to rise in the east each morning and set in the west each night. Observations made at the same time every night will show that a planet usually appears in the sky slightly to the east of its position the previous night. Periodically, a planet will appear to change direction for several nights or more and move slightly to the west of its previous position. However, the planet does not actually change direction along its orbital path. Such an apparent reversal in direction occurs whenever Earth "overtakes" an outer planet. For example, because Saturn requires almost 30 Earth years to complete one orbit around the sun, Earth often passes between Saturn and the sun. As this occurs, Saturn appears to gradually slow down and then temporarily reverse its course against the background of stars.

Mercury and Venus also have unique apparent motions. Because their orbits are smaller in diameter than Earth's, both appear to move from one side of the sun to the other. When either planet appears east of the sun to observers on Earth, it appears as an "evening star" and when west of the sun, as a "morning star."

Several images were combined to track Mars's retrograde, or "backward," motion in Earth's sky. Normally, Mars appears each night slightly to the east of its position the previous night. When Earth passes between Mars and the sun, however, Mars seems to temporarily reverse its course in the sky. The composite was put together so that each star shows up only once in the image. The small dotted line at center-right is a series of images of the planet Uranus.

Tunç Tezel/APOD/NASA

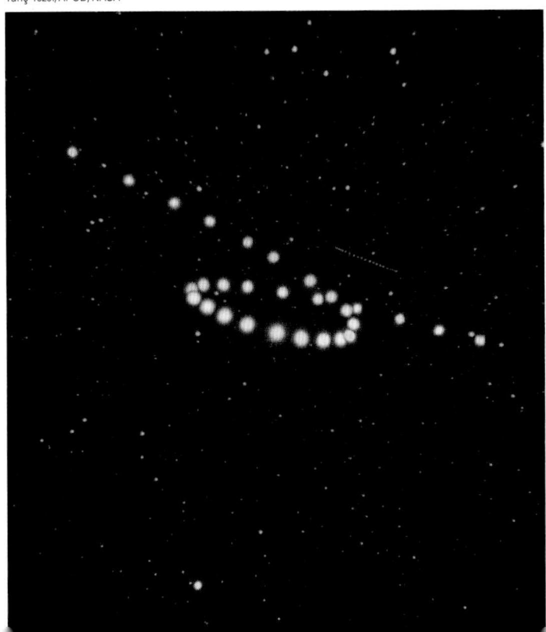

The natural satellites, or moons, revolving around the planets follow the same laws of orbital motion as do the planets. Most of the solar system's larger moons and many small moons orbit in the same direction as the planet they circle and roughly in the planet's equatorial plane (the plane that is perpendicular to the planet's rotational axis and passes through the center of the planet). These moons are called regular satellites. In addition, most large moons, including Earth's, rotate on their axes once for each revolution around the planet. As a result, these satellites always show the same side to the planet.

Formation and Evolution

Although the origin of the solar system is uncertain, most scientists believe that it began to develop about 4.6 billion years ago from a large cloud of gas and dust called the solar nebula. The gravity of the cloud began pulling the cloud's matter inward. As the cloud contracted, it began spinning faster and faster and it flattened into a disk. As the material within the cloud compressed, it grew hot. This caused the dust in the cloud to become gaseous. Most of the cloud's mass was drawn toward the center, eventually forming the sun. The planets developed from the remaining material—the disk of gas spinning around the forming sun—as it cooled. This explains why the planets orbit the sun in nearly the same plane and in the same direction.

Gases in the cooling disk condensed into solid particles, which began colliding with each other and sticking together. Larger objects began to form. As they traveled around the disk, they swept up smaller material in their paths, a process known as accretion. The larger gravity of the more massive objects also allowed them to attract more matter. Over time, much of the matter clumped together into larger bodies called planetesimals. Ultimately, they formed larger protoplanets, which developed into the planets.

The inner and outer planets developed so differently because temperatures were much hotter in the regions near the developing sun. Close to the center of the solar nebula, the material in the disk condensed into small particles of rock and metal. These particles eventually clumped together into the planetesimals that formed the rocky, dense inner planets.

Farther from the developing sun, the cooler temperatures allowed not only rock and metal to develop but also gas and the ices of such abundant substances as water, carbon dioxide, and ammonia. The availability of these ices to the forming outer planets allowed them to become much larger than the inner planets. Eventually, the outer planets grew massive enough for their gravity to be able to attract and retain even the lightest elements, hydrogen and helium. These are the most abundant elements in the universe, so the planets were able to grow enormous. They also developed compositions fairly similar to that of the sun. Each young outer planet had its own relatively cool nebula from which its regular satellites formed. The irregular satellites are generally thought to be asteroids or other objects that were captured by the planets' strong gravity.

Collisions between the forming planets and large planetesimals probably had dramatic effects. The numerous impact craters on the oldest surfaces of some inner planets and some moons are believed to have been created from such collisions. Astronomers think that Earth's moon originally may have formed from material scattered by a violent collision of Earth and a protoplanet about the size of Mars. This material may have settled into orbit around Earth and accreted to form the moon. A protoplanet also may have slammed into the developing Mercury and stripped away much of its outer rocky mantle. This would explain why Mercury's core takes up such a large percentage of the planet's interior. Other protoplanets may have crashed into Venus, greatly slowing its rotation, and Uranus, knocking the planet nearly on its side.

As the planets accreted, their interiors grew hot and melted. In a process known as differentiation, heavier materials sank to the centers, generating more heat and, in many planets, gradually forming cores. In the inner planets, the sinking of the heavier materials displaced lighter, rocky materials upward, forming mantles of rock. The most buoyant materials rose to the top and solidified into surface crust. Lighter elements escaped from the planets' interiors and formed atmospheres and, on Earth, oceans.

In addition to the heat generated by accretion and differentiation, the planets had a third source of internal heat: the decay of certain radioactive elements in their interiors. Since the planets' formation, many

Planets form from disks of gas and dust swirling around new stars. An image taken by the Hubble Space Telescope shows a disk that might be producing planets around a young star named HD 141569A. The star lies some 320 light-years from Earth. The photograph has been modified to simulate what the disk would look like if viewed from above, and false colors were added to better show the disk's structure. To reveal the disk, the star's light was blocked out, so a black central region appears in place of the star. A nearby double-star system appears at upper left.

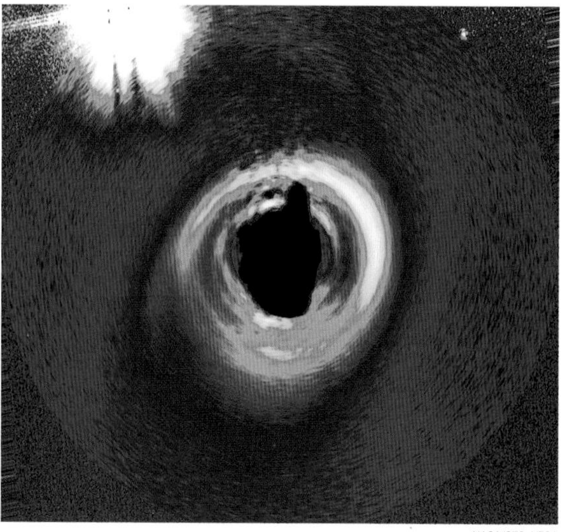

NASA/STScI/ESA

F.J. Doyle/National Space Science Data Center

Ancient craters mark the surface of the far side of Earth's moon, shown in an image taken by the Apollo 16 spacecraft. Many of the craters are the result of the moon being bombarded with asteroid-sized objects during its early history. In fact, most astronomers believe that the moon itself formed from fragments created when a larger object slammed into Earth.

ESO

What some astronomers believe to be the first direct image of an exoplanet was captured by the European Southern Observatory's Very Large Telescope in Chile. The planetlike object, which appears as the smaller, reddish object, is thought to be about five times as massive as Jupiter. It orbits a brown dwarf, which is the larger, brighter object. The distance between the two is almost twice the distance between Neptune and the sun. The image was created from three exposures taken at near-infrared wavelengths.

of their physical characteristics have been determined by the manner in which the bodies generated and lost their internal heat. For example, the release of internal heat accounts for the volcanic and tectonic activity that has shaped the crusts of the inner planets. In smaller bodies such as Mercury, Earth's moon, and many satellites of the outer planets, the internal heat escapes to the surface relatively quickly. As a result, the surface initially undergoes rapid, violent changes. Then, when most of the body's internal heat has dissipated, the surface features stabilize and remain largely undisturbed as the body ages. Larger bodies such as Earth and Venus lose their heat more slowly. The outer planets are so large that much of their internal heat is still being released.

Scientists developed these theories based on observations of the solar system. The discovery of planets outside the solar system has challenged some of the details. For example, astronomers have discovered huge gaseous planets that are closer to their stars than Mercury is to the sun. This seems to contradict the idea that huge planets can form only in the regions far from the central star. Perhaps these planets initially developed farther away from the star, or perhaps the theories about solar system formation need adjusting. The idea that solar systems develop from contracting, spinning clouds of gas and dust, however, is still believed to be correct. Astronomers have observed such disks surrounding several young stars.

Planets Outside the Solar System

People had long wondered whether stars other than the sun have planets circling them. In the 1990s astronomers found the first evidence that such planets exist. It is now known that there are numerous planets in numerous other solar systems. These planets are called extrasolar planets or exoplanets.

Such distant planets are very difficult to see directly with Earth-based telescopes. They are so far away that they would appear very small and dark, and they would normally be obscured by the glare of the stars they orbit. Also, as seen from Earth, a planet and its star are usually too close together for a telescope to optically resolve, or separate, their respective images.

The first exoplanets discovered were identified by other means. Some methods involve noting the planets' gravitational effects on the observed motion of their parent stars. For instance, a planet can be detected by the small, periodic wobbles it produces in the parent star's position in space or by deviations in the star's velocity as viewed from Earth. Another method has been used to detect exoplanets that pass directly between their stars and Earth. This causes a kind of eclipse called a transit, allowing astronomers to detect very slight dimmings of the star's apparent brightness.

Using indirect means, astronomers in the early 1990s identified three planets circling a pulsar—an extremely dense, rapidly spinning neutron star. In 1995 astronomers announced the discovery of a planet orbiting a star more like the sun, 51 Pegasi. Within 15 years of the initial discoveries, more than 200 exoplanets had been identified. In 2004–05 astronomers captured the first direct images of what appears to be a planetlike object. It orbits a brown dwarf, however, and astronomers were divided as to whether it should be called a planet. In 2008 two groups announced images of what were more widely accepted as probable exoplanets orbiting stars: visible-light images of a planet orbiting Fomalhaut and infrared images of three planets orbiting HR 8799. (*See also* Astronomy.)

MERCURY

The planet that orbits closest to the sun is Mercury. It is also the smallest of the eight planets in the solar system. These features make Mercury difficult to view from Earth, as the small planet rises and sets within about two hours of the sun. Observers on Earth can only ever see the planet during twilight, when the sun is just below the horizon. Relatively little was known about Mercury until the Mariner 10 spacecraft visited it in 1974–75. It was more than 30 years before another spacecraft, Messenger, visited the planet.

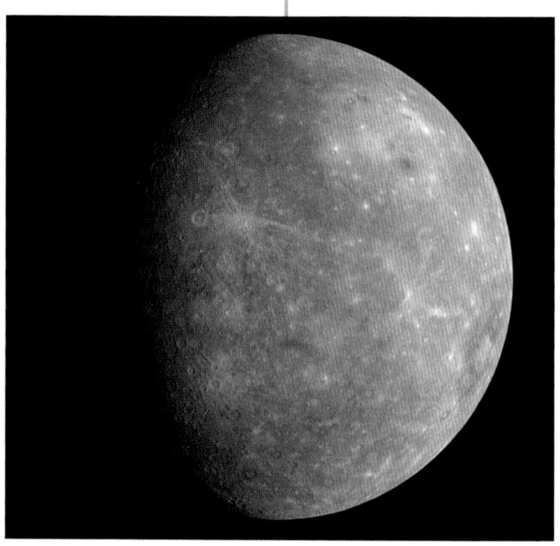

NASA/Johns Hopkins University Applied Physics Laboratory/Carnegie Institution of Washington

Numerous craters and impact basins mark the surface of Mercury. An image taken by the Messenger probe during its first flyby in January 2008 shows part of the hemisphere missed by Mariner 10 in 1974–75. The Caloris impact basin is visible as the large, circular lighter-colored area at upper right. The colors are accentuated somewhat compared to how the human eye would see them, as they include infrared data.

Basic Planetary Data

Mercury's orbit lies between the sun and the orbit of Venus. Along with Venus, Earth, and Mars, Mercury is one of the inner planets nearest to the sun. The inner planets are also known as the terrestrial, or Earth-like, planets. They are dense, rocky bodies that are much smaller than the solar system's outer planets. Mercury has no known moons.

Size, mass, and density. Mercury is the smallest planet in both mass and diameter. It is about 18 times less massive than Earth. With a diameter of about 3,032 miles (4,879 kilometers), Mercury is not quite two fifths the size of Earth. It is only about a third larger than Earth's moon. In fact, two moons in the solar system—Jupiter's moon Ganymede and Saturn's moon Titan—are larger than Mercury.

Mercury is the densest planet in the solar system, followed by Earth (if one takes into account the planets' internal compression because of gravity). Mercury is unusually dense because it is composed of a high percentage of metal. The metal is concentrated in a comparatively huge core, which accounts for nearly 75 percent of Mercury's diameter.

Appearance from Earth. Mercury can be seen from Earth without a telescope. It always appears close to the sun (within about 28 angular degrees). For this reason, the planet can only be seen near the horizon. At certain times of the year it appears as a "morning star" just before sunrise, while at other times it appears as an "evening star" just after sunset.

Because Mercury's orbit lies between Earth's orbit and the sun, Mercury displays phases like those of the moon and the planet Venus. These phases can only be seen with the aid of a telescope. Mercury sometimes looks like a crescent to observers on Earth. At other times, when Mercury and Earth are in different positions, more of the sunlight reflected off Mercury can be seen from Earth. Mercury then appears as a half or fuller disk.

About a dozen times each century, Mercury passes directly between Earth and the sun. This event, called a transit, is a type of eclipse. During a transit of Mercury, the planet appears as a small black disk against the background of the bright sun. Binoculars or a small telescope, safely equipped with a solar filter, are required to observe a transit, preferably by projection of the sun's image onto a white card. The next transit of Mercury is in 2016.

NASA/TRACE/SMEX

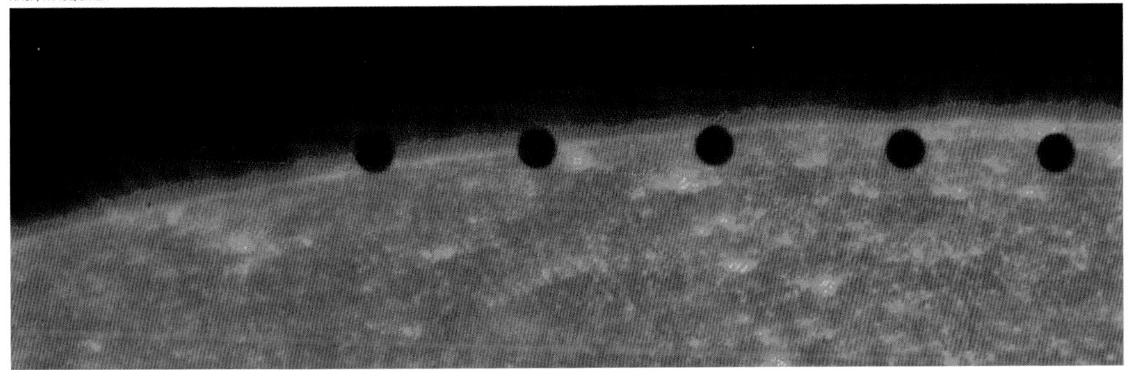

Five separate images were combined to show Mercury crossing in front of the sun during a type of eclipse called a transit on Nov. 15, 1999. The images were taken in ultraviolet light by the Transition Region and Coronal Explorer (TRACE) satellite in Earth orbit. The time between successive images was about seven minutes.

Orbit and spin. Like all the planets, Mercury travels around the sun in an elliptical (oval-shaped) orbit. Mercury's orbit is the most eccentric, or elongated, of all the planets. Its orbit is also the most tilted. The plane of Mercury's orbit is tipped about 7 degrees relative to the ecliptic, or the plane of Earth's orbit. Mercury completes one orbit around the sun about every 88 Earth days. In other words, one year on Mercury lasts some 88 Earth days.

The planet was named after the ancient Roman god Mercury, the counterpart of the ancient Greek god Hermes. Like Hermes, the swift-footed messenger of the gods, the planet Mercury is known for the speed with which it moves across the sky. The planet circles the sun at an average rate of about 30 miles (48 kilometers) per second, the fastest of the eight planets.

Facts About Mercury

Average Distance from Sun. 35,984,000 miles (57,910,000 kilometers).

Diameter at Equator. 3,032 miles (4,879 kilometers).

Average Orbital Velocity. 29.8 miles/second (47.9 kilometers/second).

Year on Mercury.* 88.0 Earth days.

Rotation Period.† 58.6 Earth days.

Day on Mercury (Solar Day). 175.9 Earth days.

Tilt (Inclination of Equator Relative to Orbital Plane). Near 0°.

General Composition. Iron core, silicate rocks.

Average Surface Temperature. 332° F (167° C).

Weight of Person Who Weighs 100 Pounds (45 Kilograms) on Earth. 37.8 pounds (17.1 kilograms).

Number of Known Moons. 0.

*Sidereal revolution period, or the time it takes the planet to revolve around the sun once, relative to the fixed (distant) stars.

†Sidereal rotation period, or the time it takes the planet to rotate about its axis once, relative to the fixed (distant) stars.

Although Mercury moves along its orbit very quickly, it spins slowly. It takes almost 59 Earth days to complete one rotation about its axis. Mercury rotates on its axis only three times for every two revolutions it makes around the sun. The combination of a slow spin and a fast orbit leads to an unusual situation. A day on Mercury—the time it takes for the sun to appear straight overhead, to set, and then to rise straight overhead again—lasts about 176 Earth days. So on Mercury a day is twice as long as a year.

This characteristic, combined with Mercury's highly eccentric orbit, creates some strange effects. The planet's distance from the sun varies greatly as it travels along its orbit. The farthest Mercury gets from the sun is about 43 million miles (70 million kilometers). At that point in the planet's orbit, an observer on Mercury would see the sun appear about twice as large as it does from Earth. The closest Mercury gets to the sun is some 29 million miles (46 million kilometers). The sun would at that point appear some three times as large as it does from Earth. Even more unusually, the sun would not seem to move steadily across Mercury's sky. Its apparent speed would change depending on the viewer's location on the planet and on the planet's distance from the sun. The sun would sometimes even appear to briefly reverse its course.

Mercury's spin axis is very nearly perpendicular, or upright, relative to its plane of orbit. By comparison, Earth's axis is tilted almost 24 degrees. This inclination is the main reason there are seasons on Earth. Because Mercury's axis is not tilted, it does not have Earth-like seasons.

Atmosphere, Surface, and Interior

Atmosphere. Unlike the other planets, Mercury has no significant atmosphere, or surrounding layers of gases. At Mercury's surface, the pressure—the force exerted by the atmosphere—is less than one trillionth that at Earth's surface. Mercury's extremely tenuous layer of gases includes atoms of helium, hydrogen, oxygen, and sodium. The gases do not remain near the planet long before the sun's heat blasts them away. They are replenished partly by the solar wind, the flow of charged particles from the sun. Other gases come from asteroids and comets and from the planet's surface.

Mercury has a magnetic field similar in form to Earth's. However, Mercury's magnetic field is much weaker, at only about 1 percent the strength of Earth's.

Temperatures on Mercury vary widely. Its closeness to the sun makes it a broiling-hot world by day, with daytime surface temperatures exceeding 800° F (430° C) at parts of the planet. Because Mercury lacks a thick atmosphere to trap heat, however, the planet cools greatly at night. The temperature can drop to about –300° F (–180° C) just before dawn.

Surface. Mercury's surface is dry and rocky. Much of it is heavily cratered, somewhat like Earth's moon. Impact craters form when meteorites, asteroids, or comets crash into a rocky planet or similar body, scarring the surface. Planetary scientists can estimate the age of a surface by the number of impact craters on it. In general, the more craters a surface has, the older it is. Mercury's heavily cratered surfaces are probably ancient.

Between the planet's heavily cratered regions are areas of flat and gently rolling plains with fewer craters. Elsewhere there are smooth, flat plains with very few craters. Volcanic lava flows probably smoothed the surfaces of these plains.

Impacts have formed craters of all different sizes. Mercury also has several huge impact basins, each of which has multiple rings in a bull's-eye pattern. The most prominent impact basin, Caloris, measures about 900 miles (1,550 kilometers) across. Along the basin's rim, mountains rise to heights of nearly 1.9 miles (3 kilometers). Images captured by the Messenger spacecraft also revealed volcanoes around the rim. On the opposite side of the planet is an area of strange hilly terrain. Scientists think it resulted from the same impact that created the Caloris basin. The crash caused seismic waves to ripple through the planet. The waves came to a focus on the part of Mercury opposite the crash site, where they warped the terrain.

Hundreds of long, steep cliffs called scarps also mark the planet's surface. Planetary scientists believe

Courtesy of John Harmon, Arecibo Observatory

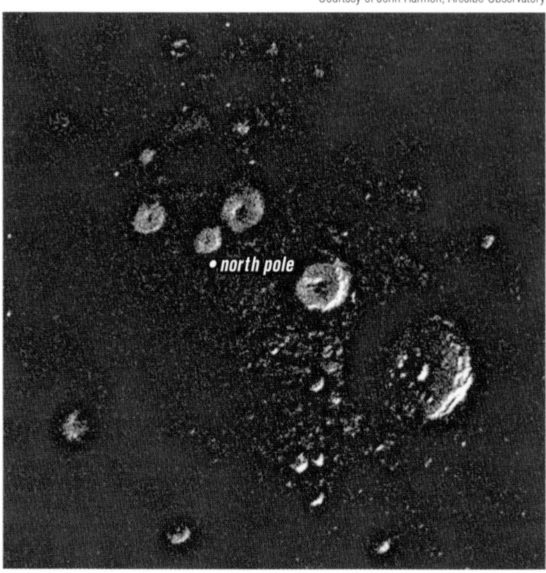

Although Mercury is the planet closest to the sun, water ice may exist near the poles in craters that are permanently in the shade. A radar image taken with the Arecibo radio telescope shows bright features in craters near the planet's north pole. Scientists believe that the bright features are probably deposits of water ice. Because Mercury's spin axis has virtually no tilt, the polar crater bottoms are never tipped toward the sun.

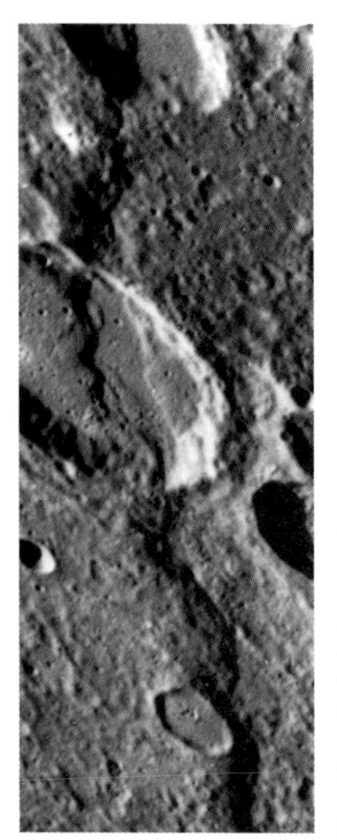

Discovery Rupes Scarp is one of many long, curving cliffs that are thought to have formed when Mercury shrank in the past. The scarp is almost 300 miles (500 kilometers) long and, in some places, more than about 0.6 mile (1 kilometer) high. The image was taken by the Mariner 10 spacecraft during its third flyby of the planet in March 1975.

NASA/JPL/Northwestern University

Mariner 10 provided the first close-up images of Mercury. Each view below is a mosaic of many images taken by the craft during its first flyby of the planet in March 1974. Each view shows about half of the hemisphere that was in sunlight at the time.

NASA/JPL

that, at some point in Mercury's history, part of the planet's interior (the mantle) began to cool and shrink. As the planet shrank, the crust buckled and cracked, forming the many scarps. Mercury may even still be shrinking.

Interior. Like Earth, Mercury has three separate layers: a metallic core at the center, a middle rocky layer called a mantle, and a thin rocky crust. In both planets, the core is made mostly of iron. However, Mercury's core is proportionally much larger than Earth's. The core takes up about 42 percent of Mercury's volume, compared with only about 16 percent for Earth. This accounts for Mercury's great density.

Observation and Exploration

Mercury has been known for at least 5,000 years. As mentioned above, its closeness to the sun makes it difficult to observe from Earth. Moreover, the Hubble Space Telescope and other Earth-orbiting instruments are too sensitive to be pointed that close to the sun. Astronomers have used radar to study Mercury by sending radio waves toward the planet and detecting and measuring the waves that bounce back.

The planet's nearness to the sun also presents challenges for space probes, which must contend with great heat and the enormous pull of the sun's gravity. A spacecraft needs a lot of energy in order to enter into orbit around Mercury.

Much of the information known about Mercury comes from images and data transmitted by the Mariner 10 spacecraft, the first to visit the planet. The National Aeronautics and Space Administration (NASA) launched the craft in November 1973 toward Venus for the initial leg of its mission. Mariner 10 became the first spacecraft to use a "gravity assist," drawing on Venus' gravitational field to boost its speed and divert its course toward Mercury. It captured the first close-up photographs of Mercury in March 1974. Mariner 10 encountered Mercury twice more. Its final and closest pass, in March 1975, brought it to within about 200 miles (325 kilometers) of the surface.

Mariner 10's orbital path around the sun allowed it to photograph only one side of Mercury. Low-resolution radar images taken from Earth suggested that the planet's other hemisphere has broadly similar terrain; this was confirmed when the Messenger spacecraft photographed areas of the surface unseen by Mariner. NASA launched Messenger, only the second spacecraft ever sent to Mercury, in August 2004. It was designed to be the first probe to orbit the planet, getting gravity assists during a flyby of Earth, two flybys of Venus, and three flybys of Mercury.

NASA

NASA technicians carefully lift the Messenger spacecraft in order to move it to a prelaunch testing stand. It was launched on Aug. 3, 2004.

Exploring Mercury: Spacecraft Missions

Spacecraft	Type of Mission	Launch Date	Country or Agency	Highlights
Mariner 10	flyby	Nov. 3, 1973	U.S.	First mission to Mercury. Flew by Venus once (1974) before flying by Mercury three times (March 29, 1974; Sept. 21, 1974; March 16, 1975). Transmitted first close-up images of Mercury's surface plus data about its atmosphere and magnetic field. First spacecraft to use gravity assist.
Messenger	orbiter	Aug. 3, 2004	U.S.	Second mission to Mercury. Designed to use several gravity assists (one flyby of Earth, two flybys of Venus, and three flybys of Mercury) before entering orbit around Mercury.

NASA/JPL/Caltech (NASA photo # PIA00271)

Scientists use radar to pierce the thick clouds shrouding Venus and "see" the surface below. An image generated by computer from radar data collected by the Magellan spacecraft shows the surface of the northern hemisphere. Maxwell Montes, Venus' highest mountain range, appears as a bright spot just below the center. The range is about the size of the Himalayas on Earth. The colors were added to the image to simulate those observed at the surface by Venera landers.

VENUS

The second planet from the sun is Venus. After the moon, Venus is the most brilliant natural object in the nighttime sky. It is the closest planet to Earth, and it is also the most similar to Earth in size, mass, volume, and density. These similarities suggest that the two planets may have had similar histories. Scientists are thus intrigued by the question of why Venus and Earth are now so different.

Venus was named after the ancient Roman goddess of love and beauty, but its conditions are anything but hospitable and inviting to humans. Unlike Earth, Venus is extremely hot and dry. The planet is always shrouded by a thick layer of clouds. Venus has a massive atmosphere, or surrounding layers of gases, composed mainly of carbon dioxide. This thick atmosphere traps heat, making Venus the hottest planet in the solar system.

The permanent blanket of clouds also makes it difficult to study the planet. Little was known about the surface and atmosphere until the 1960s, when astronomers made the first radar observations of Venus and unmanned spacecraft began visiting the planet.

Basic Planetary Data

Venus' orbit lies between the orbits of Mercury and Earth. These three planets plus Mars—the four innermost planets in the solar system—are known as the terrestrial, or Earth-like, planets. They are all fairly dense and rocky, with solid surfaces. Like Mercury, Venus has no known moon.

Size, mass, and density. Venus is the third smallest planet in the solar system, after Mercury and Mars. It is a near twin of Earth in size, mass, and density. Venus' diameter is about 7,521 miles (12,104 kilometers), compared with some 7,926 miles (12,756 kilometers) for Earth. Its mass is approximately 80 percent of Earth's, and its density is about 95 percent of Earth's. The surface gravity of the two planets is also of similar strength.

Appearance from Earth. Along with Mercury, Venus is an "inferior" planet, or one whose orbit is smaller than Earth's. For this reason, Venus always appears in Earth's sky in roughly the same direction as the sun. At some times of the year the planet can be seen as a "morning star," appearing in the hours before sunrise. At other times it can be seen as an "evening star" in the hours after sunset. Venus often can be seen in clear skies during daylight, if the observer knows exactly where to look.

Because Venus orbits closer to the sun than Earth does, it exhibits phase changes as viewed from Earth. These phases are similar to those of the moon and Mercury. Venus sometimes appears as a thin crescent and sometimes as a half or fuller disk. It passes through one cycle of phases about every 584 Earth days. The phases can be seen easily in high-power binoculars or a small telescope.

Facts About Venus

Average Distance from Sun. 67,232,000 miles (108,200,000 kilometers).

Diameter at Equator. 7,521 miles (12,104 kilometers).

Average Orbital Velocity. 21.7 miles/second (35.0 kilometers/second).

Year on Venus.* 224.7 Earth days.

Rotation Period† 243 Earth days (retrograde).

Day on Venus (Solar Day). 116.8 Earth days.

Tilt (Inclination of Equator Relative to Orbital Plane). 177°.

Atmospheric Composition. 96% carbon dioxide; 3.5% molecular nitrogen; 0.02% water; trace amounts of carbon monoxide, molecular oxygen, sulfur dioxide, hydrogen chloride, other gases.

General Composition. Iron core, silicate rocks.

Average Surface Temperature. 867° F (464° C).

Weight of Person Who Weighs 100 pounds (45 kilograms) on Earth. 87.8 pounds (39.8 kilograms).

Number of Known Moons. 0.

*Sidereal revolution period, or the time it takes the planet to revolve around the sun once, relative to the fixed (distant) stars.

†Sidereal rotation period, or the time it takes the planet to rotate about its axis once, relative to the fixed (distant) stars.

Venus rarely but regularly passes directly between Earth and the sun. This event is a type of eclipse called a transit. The planet then appears to observers on Earth as a small black disk crossing the bright disk of the sun. Two transits of Venus occur about every 125 years (*see* Transits of Venus on page 31). The transits occur in pairs eight years apart. One should never look at a transit without special protective equipment. Magnification provides the best views. For safe viewing, binoculars or a telescope can be outfitted with special solar filters or used to project the image of the transit onto white cardboard.

Orbit and spin. All eight planets revolve around the sun in elliptical, or oval-shaped, orbits. Venus' orbit is the most nearly circular of all the planets. It orbits the sun at a mean distance of about 67 million miles (108 million kilometers). This is about 30 percent closer to the sun than that of Earth's orbit. At its closest approach to Earth, Venus is about 26 million miles (42 million kilometers) away; at its farthest, Venus is some 160 million miles (257 million kilometers) away. Venus completes one orbital revolution about every 225 Earth days, which is the length of one year on Venus.

Venus' rotation is unusual in a couple of ways. It spins about its axis very slowly, completing one rotation about every 243 Earth days. It is the only planet in the solar system that takes longer to rotate once about its axis than to travel once around the sun. These two motions combine so that a day on Venus—the time it takes for the sun to appear straight overhead, to set, and then to rise straight overhead again—lasts about 117 Earth days.

Because of its slow rotation, Venus is more nearly spherical than Earth and most other planets. The force from a planet's rotation generally causes some bulging at the equator and flattening at the poles. These distortions are minimized on Venus.

Venus also rotates in retrograde motion, or the direction opposite that of most of the other planets and members of the solar system. Six of the eight planets rotate clockwise when viewed from above the northern pole, while only Venus and Uranus rotate counterclockwise. To an observer on Venus, the sun would appear to rise in the west and set in the east (if one could see through the thick clouds). Venus' spin axis is tilted only about 3 degrees relative to the plane of its orbit. This means that seasonal variations on the planet are probably very slight.

Unlike Earth and most of the other planets, Venus does not have a global magnetic field. This might result from its extremely slow rotation rate. Scientists think that a planet's rotation helps drive the fluid motions in the planet's core that generate a magnetic field.

Atmosphere, Surface, and Interior

Atmosphere and greenhouse effect. Venus has by far the most massive atmosphere of the four terrestrial planets. The pressure exerted by the atmosphere at the planet's surface is about 95 bars, or 95 times the atmospheric pressure at sea level on Earth. It is composed of more than 96 percent carbon dioxide and about 3.5 percent molecular nitrogen, with only trace amounts of other gases.

The layer of clouds that perpetually blankets Venus is very thick. The main cloud deck rises from an

Large V-shaped bands in Venus' clouds are revealed in a photograph taken in ultraviolet light by the Pioneer Venus 1 spacecraft. Although the planet's cloud cover is nearly featureless in visible light, ultraviolet imaging shows a distinctive structure and pattern. Color was added to the image to emulate Venus' yellow-white appearance to the eye.

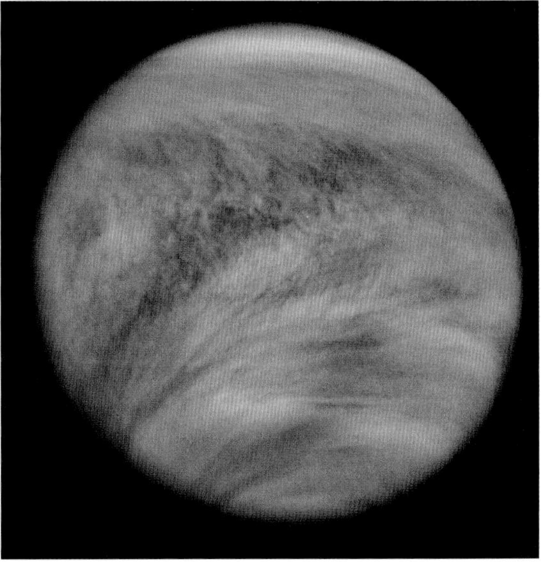

NASA/JPL

altitude of about 30 miles (48 kilometers) to nearly 42 miles (68 kilometers). In addition, thin hazes extend several miles above and below the main deck. The clouds are made of microscopic particles, predominantly droplets of sulfuric acid. The clouds may also contain solid crystals. Some cloud-top regions appear dark in ultraviolet light. This might indicate the presence of sulfur dioxide, chlorine, or solid sulfur.

Although Venus rotates slowly (once every 243 days), its atmosphere circulates surprisingly rapidly. At cloud level the atmosphere completely circles the planet every four days. Winds blow at some 220 miles (360 kilometers) per hour at the cloud tops. The wind velocity decreases greatly with altitude, and at the surface the winds are quite slow.

Even though Venus is closer to the sun than Earth is, Venus absorbs less sunlight than Earth does. Venus' thick clouds allow only a little light through. About 85 percent of the sunlight that strikes the clouds gets reflected back into space.

The sunlight that does penetrate the clouds is absorbed by the lower atmosphere and surface. As the light heats the lower atmosphere and the ground, they radiate some of the energy back at longer, infrared wavelengths. On Earth most such energy escapes back into space. This keeps Earth's surface relatively cool. On Venus, however, the thick atmosphere traps much of the reradiated infrared energy. Energy at infrared wavelengths cannot pass through a carbon dioxide-

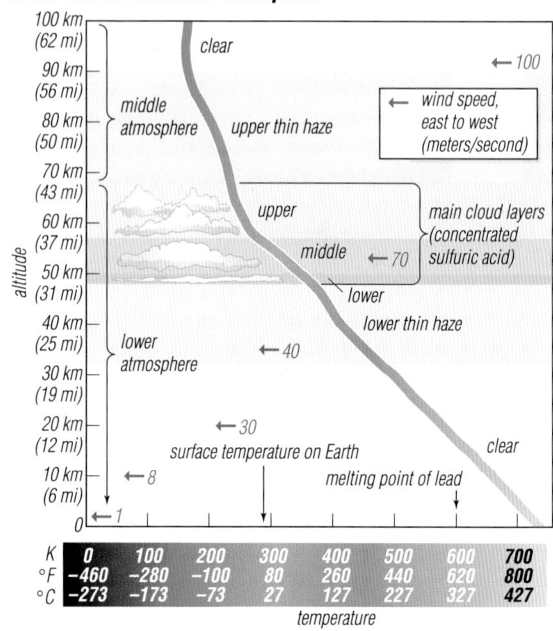

A graph shows temperatures and pressures at different altitudes in Venus' lower and middle atmosphere, based on measurements made by the Pioneer Venus mission's atmospheric probes and other spacecrafts. Red arrows indicate wind speeds, ranging from about 100 meters per second (225 miles per hour) near the top of the middle atmosphere to only a slight breeze at the surface.

A surface map of Venus shows the planet's global topography. The map is color coded according to the key at right so that the lowest elevations are in blue and the highest are in red. The elevation figures are expressed as distance from the center of the planet, ranging from 6,048 to 6,064 kilometers (3,758 to 3,768 miles). Selected major surface features and spacecraft landing sites are labeled. This Mercator projection was derived from laser altimetry data collected by the Magellan spacecraft.

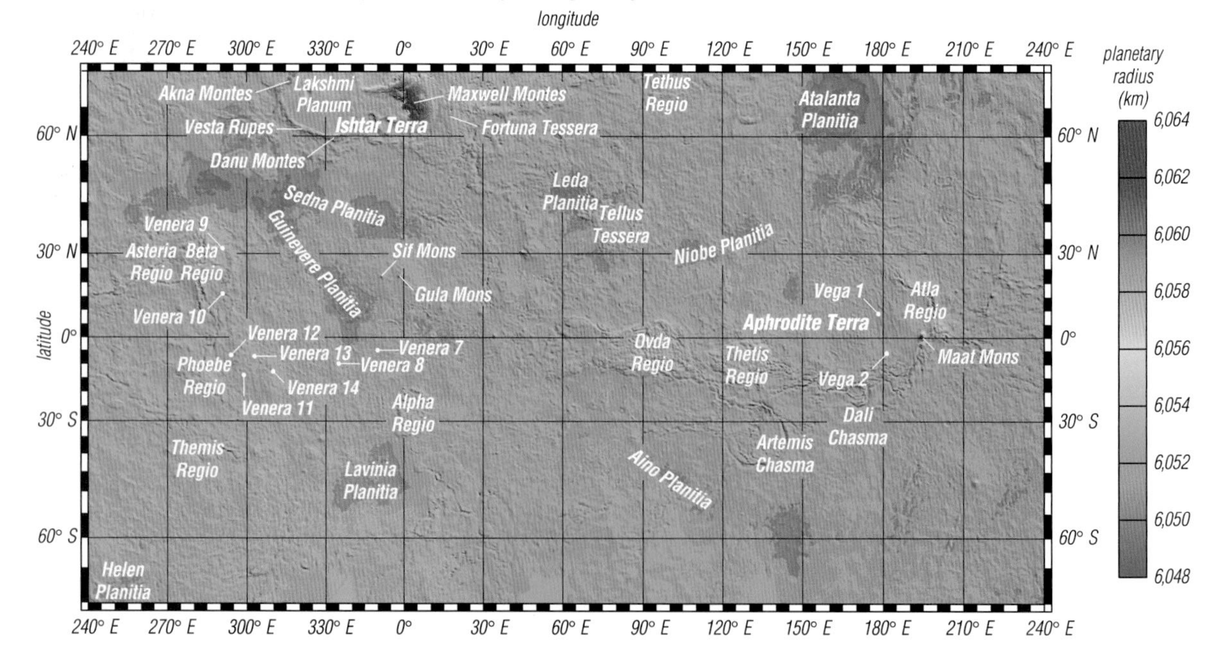

Source: NASA/JPL/Massachusetts Institute of Technology

Courtesy of C.M. Pieters through the Brown/Vernadsky Institute to Institute Agreement and the U.S.S.R. Academy of Sciences, and C.M. Pieters et al. "The Color of the Surface of Venus," *Science*, vol. 234, p. 1382, Dec. 12, 1986, copyright © 1986 by the American Association for the Advancement of Science

Flat rock slabs and soil on the surface of Venus appear in panoramic 170° images taken by the Venera 13 lander in natural light (top) and color corrected (bottom). The planet's clouds and thick atmosphere filter out blue light, which gives the natural-light image an orange hue. The color-corrected image simulates what the surface would look like in white light (without the blue filtered out). The spacecraft itself is visible at the bottom of the images.

rich atmosphere as easily as shorter-wavelength visible light can.

This phenomenon, called the greenhouse effect, makes Venus extremely hot. The planet's surface temperatures reach 867° F (464° C), which is hot enough to melt lead. Venus is even hotter than Mercury, the planet closest to the sun. The rocks on Venus may glow faintly red from their own heat.

Studying the greenhouse effect on Venus has given scientists an improved understanding of the more subtle but very important influence of greenhouse gases in Earth's atmosphere. (Rising levels of carbon dioxide, methane, and other so-called greenhouse gases in Earth's atmosphere are thought to be causing global warming on Earth.)

Surface. Venus has a dry, rocky surface. In the 1970s and '80s the Soviet Union's Venera series of unmanned spacecraft obtained the first detailed information about the planet's surface. Photographs taken by robotic Venera landers revealed plains strewn with flat, slabby rocks and a darker, fine-grained soil. Venera landers also measured the chemical composition of the surface at the landing sites. Their analysis suggested that the rock composition might be similar to basalts found on the ocean floors of Earth.

Earth-based observatories and several orbiting spacecraft have mapped Venus' surface using radar. The radar maps reveal diverse and geologically complex surface terrain. Most of the planet consists of gently rolling plains. There are also several lowland areas and two continent-sized highlands: Ishtar Terra and Aphrodite Terra. Ishtar is about the size of Australia, while Aphrodite is roughly the size of South America. Ishtar has a central plateau surrounded by mountains, including the enormous Maxwell Montes range. Its peaks rise to the highest elevations on Venus, about 7 miles (11 kilometers) above the planet's average surface elevation.

As on Earth, geologic activity has shaped the surface terrain. Upward and downward movements within Venus' outer shell have folded, fractured, and otherwise deformed the crust. Features that probably

A shield volcano on Venus named Sif Mons appears in a computer-generated image based on radar data from the Magellan spacecraft. The length of the lava flows suggests that the lava was very fluid. The volcano is about 1.2 miles (2 kilometers) high, and its base is about 200 miles (300 kilometers) in diameter. The image is somewhat exaggerated in the vertical direction to accentuate the relief. Color was added to simulate the appearance of the surface in photographs taken by Venera landers.

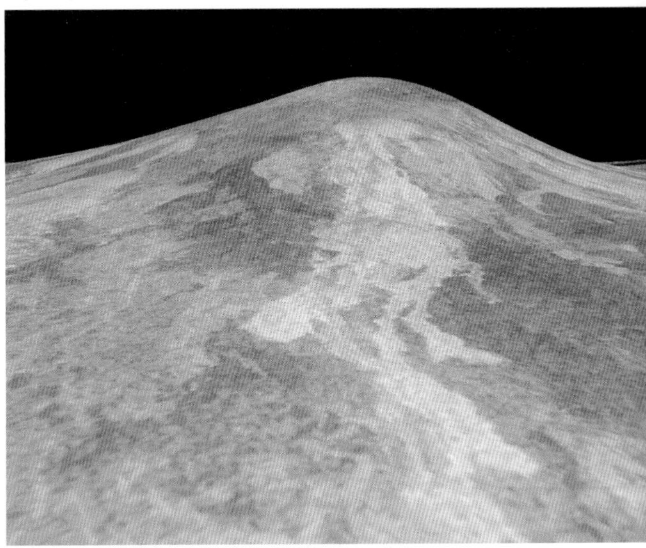

NASA/JPL

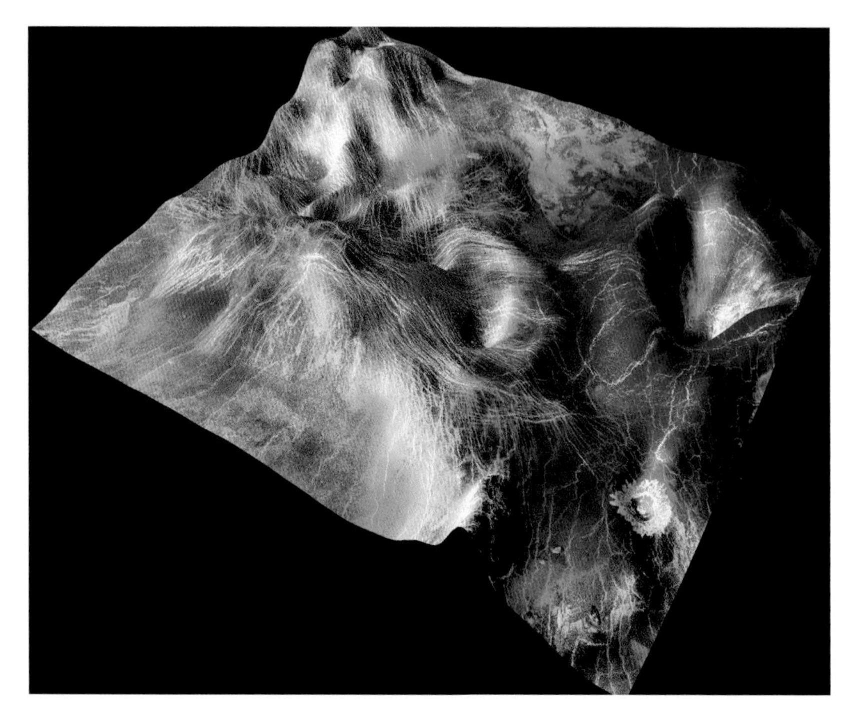

Venus' complex surface features include coronae, which are sets of faults, fractures, and ridges in circular patterns. A false-color, computer-generated image based on radar data from the Magellan spacecraft shows a corona in an early stage, at left, characterized by raised crust, and a corona in a later stage, at right, in which the center has begun to sag. The image is highly exaggerated in its vertical direction. The topography has been color coded to indicate differences in the radio waves and heat emitted by the surface; these emissions can provide information about surface composition.

NASA/JPL/Caltech (NASA photo # PIA00307)

formed this way include mountain belts and rift valleys, or deep, narrow troughs. Another such type of terrain on Venus, called tessera, is very rugged, complex, and deformed. Tessera terrain typically has sets of parallel troughs and ridges that cut across one another at a wide range of angles. Hundreds of features called coronae are also found on the surface. A corona is a set of faults, fractures, and ridges in a circular or oval pattern. Some have a raised outer rim and a sagging center. Coronae probably form when blobs of molten material rise up in the planet's interior and distort the crust.

Many of Venus' surface features are associated with volcanic activity. The planet has more than a hundred shield volcanoes, and enormous fields of lava flows cover most of the rolling plains. There are also numerous small volcanic cones.

Like other planets, Venus has impact craters, which form when asteroids crash into the surface. However, Venus does not have craters smaller than about a mile (1.5 kilometers). This is because the planet's thick atmosphere slows down and breaks apart smaller asteroids. Scientists can estimate the age of a solid planet's surface in part by analyzing its craters. In general, the more craters a surface has, the older it is. Venus has comparatively few craters, and they are randomly distributed over the surface. This indicates that Venus' surface is in all places young for a planet. Scientists believe that Venus underwent an intense period of global resurfacing only roughly 500 million years ago. One possible explanation is that the planet's outer shell may have slowly thickened until eventually it collapsed. This may have been a single event or the last in a cycle of

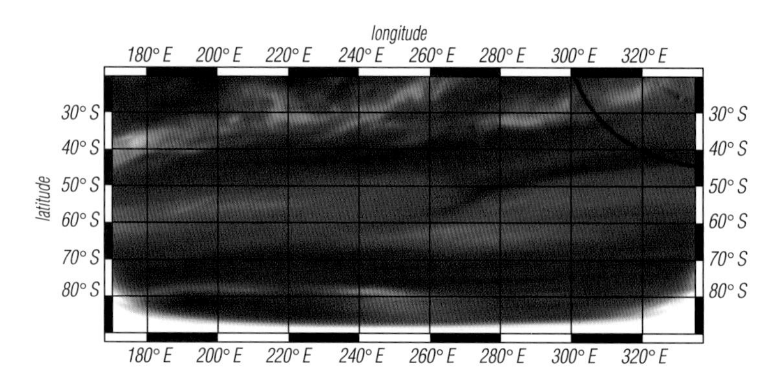

An infrared image taken by the Venus Express orbiter shows wind-blown clouds in the night sky above Venus.

Source: ESA/VIRTIS/INAF-IASF/Obs. de Paris-LEIA

global convulsions that each time renewed the surface.

Interior. What little is known about Venus' interior is mostly inferred from its similarity to Earth in terms of density and size. Planetary scientists theorize that Venus probably developed an interior roughly like that of Earth, with a metallic core and a rocky mantle and crust.

Venus' core probably extends outward about 1,860 miles (3,000 kilometers) from the planet's center. It likely contains iron and nickel like Earth's core. Venus' core probably also includes a less dense substance such as sulfur. Unlike most of the other planets, however, Venus has no magnetic field, so there is no direct evidence for a metallic core.

The mantle makes up the bulk of the planet. Gravitational data suggest that the crust is typically about 12–30 miles (20–50 kilometers) thick. It likely contains much basalt. As mentioned above, movements within Venus' mantle are thought to deform the crust. These movements are mainly vertical. Venus does not now seem to experience plate tectonics as Earth does (*see* Earth, "Plate tectonics"). Plate tectonics involves mainly horizontal movements of a planet's crust and upper mantle.

Tass/Sovfoto

The Soviet Venera 4 atmospheric probe, shown on Earth before its launch, took direct measurements of Venus' atmosphere in 1967. It was the first object made by humans to enter the atmosphere of another planet and transmit data back to Earth.

Observation and Exploration

Venus was observed from Earth for centuries before the invention of modern astronomical instruments. The Babylonians recorded its appearances in about 3000 BC, and ancient civilizations in China, Central America, Egypt, and Greece also observed the planet.

Telescopic observation. In the 17th century Galileo made the first telescopic observations of Venus. In 1610 he discovered the planet's phases. If Earth lay at the center of the solar system, as was then widely believed, Venus would not display such phases. Galileo's discovery was the first direct observational evidence to support Nicolaus Copernicus' then-controversial theory that Earth and the other planets orbit the sun.

Important telescopic observations were later made during the transits of Venus in the 1700s and 1800s.

Transits of Venus

Transits of Venus, a kind of eclipse in which Venus passes directly between Earth and the sun, are very rare. Below are the dates of all the transits of Venus from the first such event observed after the invention of the telescope through the 24th century. The transits occur in pairs eight years apart. See *text for information about safe viewing.*

Dec. 7, 1631	Dec. 4, 1639
June 6, 1761	June 3, 1769
Dec. 9, 1874	Dec. 6, 1882
June 8, 2004	June 6, 2012
Dec. 11, 2117	Dec. 8, 2125
June 11, 2247	June 9, 2255
Dec. 13, 2360	Dec. 10, 2368

Studying the transits of 1761 and 1769 helped astronomers determine a more accurate value for the distance between Earth and the sun. (This distance, called the astronomical unit [AU], is a basic unit for measuring distances in astronomy.)

When viewed through an optical (visible-light) telescope, Venus appears yellow-white and fairly featureless because of its permanent veil of clouds. Since the early 20th century astronomers have used wavelengths of light that lie outside the visible spectrum to uncover features of the planet that are otherwise hidden. Ultraviolet rays reveal swirls, v-shaped bands, and distinctive bright and dark markings in the clouds. Astronomers have used infrared radiation to study the composition of the atmosphere and clouds. Microwave studies have revealed the high temperatures at the planet's surface.

Radar studies have been particularly important in helping astronomers "see" through the planet's clouds. A large radio telescope outfitted with a transmitter can send out radio waves that pierce Venus' cloud screen and bounce off the planet's surface. The telescope then detects the returning radio waves.

Spacecraft exploration. Exploration of the planet with spacecraft began in the 1960s. More than 20 unmanned spacecraft have visited Venus, including craft that have flown by, orbited, and landed on the planet and that have sent probes parachuting through its atmosphere. Like Earth-based telescopes, spacecraft flying near Venus use radar to penetrate the deck of clouds and map the surface below. Missions to the surface have also had to contend with the planet's extremely high temperatures and pressures.

Exploring Venus: Major Spacecraft Missions*

Spacecraft	Type of Mission	Launch Date	Country or Agency	Highlights
Mariner 2	flyby	Aug. 27, 1962	U.S.	First mission to successfully fly near another planet and return data. Reached Venus on Dec. 14, 1962. Was backup for Mariner 1, which failed on launch.
Venera 4	flyby and atmospheric probe	June 12, 1967	U.S.S.R.	First human-made object to travel through atmosphere of another planet and return data. Probe parachuted through Venus' atmosphere on Oct. 18, 1967, analyzing its chemical composition.
Venera 7	lander	Aug. 17, 1970	U.S.S.R.	First successful soft landing on another planet. Landed on Venus on Dec. 15, 1970, returning data during descent and for 23 minutes on surface.
Mariner 10	flyby	Nov. 3, 1973	U.S.	First spacecraft to use gravity assist, flying by Venus on Feb. 5, 1974, before traveling to Mercury (1974–75). Returned first close-up ultraviolet photographs of Venus' atmosphere.
Venera 9	orbiter and lander	June 8, 1975	U.S.S.R.	Orbiter was first craft to orbit Venus. Lander returned first images from surface of another planet. Lander reached surface of Venus on Oct. 22, 1975, surviving there for 53 minutes.
Venera 10	orbiter and lander	June 14, 1975	U.S.S.R.	Landed on surface on Oct. 25, 1975, surviving 65 minutes there. Twin of Venera 9.
Pioneer Venus 1	orbiter	May 20, 1978	U.S.	Began orbiting Venus on Dec. 4, 1978. Mapped more than 90 percent of surface with radar. Returned data from several instruments for more than 14 years. Ran out of fuel in 1992 and burned up in atmosphere.
Pioneer Venus 2	atmospheric probes	Aug. 8, 1978	U.S.	Multiprobe released four atmospheric probes (one large, three small) at different regions of planet. Entered atmosphere on Dec. 9, 1978, returning data from all levels of atmosphere to surface.
Venera 13	orbiter and lander	Oct. 30, 1981	U.S.S.R.	Landed on planet on March 1, 1982, surviving on surface for 127 minutes. Analyzed nonradioactive elements in surface rocks, finding them similar to earthly basalts. Returned first color images from surface, revealing rocky landscapes bathed in yellow-orange sunlight.
Venera 14	orbiter and lander	Nov. 4, 1981	U.S.S.R.	Landed on Venus on March 5, 1982, surviving on surface for 57 minutes. Twin of Venera 13.
Venera 15	orbiter	June 2, 1983	U.S.S.R.	First craft to use high-resolution imaging radar system at another planet. Arrived at Venus on Oct. 10, 1983. Produced high-quality map of northern quarter of surface.
Venera 16	orbiter	June 7, 1983	U.S.S.R.	Arrived at Venus on Oct. 11, 1983. Twin of Venera 15.
Magellan	orbiter	May 4, 1989	U.S.	Began orbiting Venus on Aug. 10, 1990. Produced high-quality radar images of about 98 percent of planet's surface. Made first detailed map of Venus' gravitational field. Returned data for more than four Earth years.
Venus Express	orbiter	Nov. 9, 2005	ESA	First European mission to Venus. Arrived at planet on April 11, 2006, to study atmosphere. Returned first images of cloud structures over south pole.

*More than 20 spacecraft have visited Venus. This table presents some of the more notable missions.

Photo of Venus Express aboard launch vehicle courtesy of ESA/Starem—S. Corvaja

The United States Mariner and Soviet Venera missions were the first to successfully visit Venus. The National Aeronautics and Space Administration (NASA) launched the Mariner 2 spacecraft in August 1962. When it reached Venus a few months later, it became the first spacecraft to fly near another planet and return data. Approaching Venus within about 22,000 miles (35,000 kilometers), the craft surveyed the atmosphere and collected data about the planet's rotation and high surface temperatures and pressures. It found no evidence of a global magnetic field. In 1967 Mariner 5 flew closer to Venus, passing within about 2,500 miles (4,000 kilometers). Its more sensitive instruments returned more precise data about the atmosphere, including the heavy concentration of carbon dioxide. Mariner 10, the last craft of the series, took some 4,000 photographs of Venus during a flyby in 1974. It captured the first close-up ultraviolet images of Venus' clouds.

Another early NASA mission to the planet, called Pioneer Venus, included two craft. Pioneer Venus 1 and Pioneer Venus 2 arrived at the planet in 1978. The first of these spacecraft orbited Venus for several years, collecting comprehensive data on the atmosphere. Its radar instrument produced the first high-quality map of Venus' surface topography. Pioneer Venus 2, known as the Multiprobe, released four probes to collect data at different points in the planet's atmosphere.

Many Soviet missions targeted Venus from the 1960s to the 1980s. Several of the early missions failed, but the later Venera missions were outstanding successes. In 1961 Venera 1 became the first spacecraft to fly by Venus. However, mission scientists lost contact with the craft long before it reached the planet. In 1966 Venera 2 flew by Venus and Venera 3 crash-landed onto the surface. The systems aboard both craft failed and they returned no data.

Venera 4 successfully flew by Venus in 1967 and released the first probe to enter its atmosphere. The probe took the first direct measurements of the chemical composition of the planet's atmospheric gases. Veneras 5 and 6 also successfully probed the atmosphere, in 1969. In 1970 Venera 7 became the first spacecraft to land on the surface of another planet and send data back to Earth. Venera 8 included an atmospheric probe and lander, which reached the planet in 1972.

In 1975 Veneras 9 and 10, the first craft to orbit Venus, sent back the first close-up photographs of the surface. In 1978 the Venera 11 and 12 landers took detailed chemical measurements of the atmosphere on their way to soft landings. The Venera 13 and 14 landers analyzed rock samples in 1982. The radar mappers aboard the final craft of the mission, Veneras 15 and 16, produced high-quality images of the surface in 1983. Many of the types of geologic features found on Venus were first revealed by these twin Venera orbiters.

The Soviet Union sent two more craft, Vegas 1 and 2, to Venus. In 1985 each of the twin craft delivered a lander to the surface and dropped an instrument-equipped balloon into the atmosphere. (Both spacecraft later flew by Halley's comet.)

NASA/JPL

The Magellan spacecraft and an attached rocket are released into a temporary orbit around Earth from the payload bay of the space shuttle* Atlantis *in 1989. Shortly afterward, the rocket propelled Magellan on a course that took it around the sun and then into orbit around Venus.

In 1989 NASA's Magellan became the first planetary spacecraft to be launched from a space shuttle. The craft surveyed Venus from August 1990 through October 1994. Its orbit carried it around the planet every three hours while it mapped the cloud-shrouded surface in great detail. Its radar system could produce images with a resolution greater than 330 feet (100 meters). Magellan found no evidence of plate tectonics on Venus, but it revealed data that suggest the planet is still geologically active at a couple of hot spots.

The first European mission to Venus was the orbiter Venus Express. The European Space Agency (ESA) launched the craft in 2005. It began orbiting the planet in 2006, with a camera, a visible-light and infrared imaging spectrometer, and other instruments on board to study the planet's magnetic field, plasma environment, atmosphere, and surface.

In addition to missions to Venus, several spacecraft have flown past Venus while on their way to other main targets. These flybys were designed as "gravity assists," which transfer momentum from the planet to the spacecraft in order to increase the craft's velocity and adjust its course. Such gravity assists also allow spacecraft to investigate Venus while flying by. The first craft to use a gravity assist was Mariner 10, which flew past Venus in 1974 on its way to Mercury. Others have included NASA's Galileo, which flew by Venus in 1990 on its way to Jupiter, and NASA's Cassini, which flew by Venus in 1998 and 1999 on its way to Saturn.

EARTH

NASA

Astronauts aboard the Apollo 17 spacecraft captured a stunning image of Earth as the spacecraft headed to the moon in 1972. Vast oceans and seas surround the continent of Africa, the island of Madagascar, and, at top, the Arabian Peninsula. The swirling clouds in the atmosphere are concentrated above the south polar ice cap.

The third planet from the sun is Earth, our home. It contains the only places in the universe known to harbor life of any kind. Liquid water, which is essential for all known forms of life, is found in abundance on Earth. Deep, salty oceans cover more than two thirds of the surface. Also, the planet's oxygen-rich atmosphere is unique. This is actually not surprising, since large amounts of oxygen exist in the atmosphere only because living things constantly supply it. Earth's green plants take in carbon dioxide and give off oxygen, which humans and other animals need to breathe.

Earth has one natural satellite, the moon. Because of its nearness to Earth, it is the brightest object in the sky after the sun. (The moon, of course, does not generate its own light but reflects the light of the sun.) Earth's moon is large, but it is not the largest moon in the solar system. However, no other moon is as massive compared to its planet as Earth's moon. The moon's large comparative size means that it affects Earth to an unusual degree. Its influence is most evident in the ocean tides, which are caused by the pull of the moon's gravity.

Basic Planetary Data

Earth's planetary neighbors are Venus, which orbits closer to the sun, and Mars, which orbits farther from the sun. Along with Mercury, Venus, and Mars, Earth is one of the four inner, terrestrial planets, which are rocky worlds with solid surfaces. They are much denser and smaller than the four outer planets.

Size, mass, and density. Earth is the largest of the four terrestrial planets and the fifth largest of all eight planets. Its diameter is more than 2.5 times as big as that of the smallest planet, Mercury, but more than 11 times smaller than that of the largest, Jupiter.

For several hundred years almost everyone has accepted the fact that Earth is round. Actually, Earth is nearly, but not exactly, spherical. Like other planets, Earth has a slight bulge around its equator, which results from its rotation about its spin axis. Measured at sea level, the diameter of Earth around the equator is 7,926 miles (12,756 kilometers). The distance from the North Pole to the South Pole, also measured at sea level, is 7,900 miles (12,714 kilometers). Compared to the overall diameter, the difference—only about 26 miles (42 kilometers)—seems small. But compared to the height of Earth's surface features, it is fairly large. For example, the planet's tallest mountain, Mount Everest, juts less than 6 miles (9 kilometers) above sea level. Earth's total surface area is about 196,938,000 square miles (510,066,000 square kilometers).

Earth is the most massive of the terrestrial planets. Its mass is 6.587×10^{21} tons (5.976×10^{24} kilograms). In numerals this would read 6 sextillion, 587 quintillion tons. Venus has roughly 80 percent of Earth's mass; Mars, 11 percent; and Mercury, only 5.5 percent. The outer planets are much more massive than Earth, however, ranging from 14.5 times as massive (Uranus) to 320 times (Jupiter).

Earth has the greatest average density of the planets (though Mercury is denser if internal compression

Facts About Earth

Average Distance from Sun. 92,956,000 miles (149,600,000 kilometers).

Diameter at Equator. 7,926 miles (12,756 kilometers).

Average Orbital Velocity. 18.5 miles/second (29.8 kilometers/second).

Year on Earth.* 365.3 Earth days.

Rotation Period.† 23.9 Earth hours.

Day on Earth (Solar Day). 24.1 Earth hours.

Tilt (Inclination of Equator Relative to Orbital Plane). 23.5°.

Atmospheric Composition. 78% molecular nitrogen; 21% molecular oxygen; 0.93% argon; variable amounts of water vapor, carbon dioxide, ozone, and other gases; small amounts of neon, helium, methane, krypton, and other gases.

General Composition. Iron core, silicate rocks.

Average Surface Temperature. 59° F (15° C).

Number of Known Moons. 1.

*Sidereal revolution period, or the time it takes the planet to revolve around the sun once, relative to the fixed (distant) stars.

†Sidereal rotation period, or the time it takes the planet to rotate about its axis once, relative to the fixed (distant) stars.

factors are considered). Its average density is some 3.2 ounces per cubic inch (5.5 grams per cubic centimeter). This falls between the density of iron, which is denser, and that of the volcanic rock basalt, which is less dense. However, Earth is made up of many different kinds of materials with varying densities. In general, the planet's density increases with depth. The material on the continents is on average only about half as dense as Earth's average density. The density of the core is thought to be roughly twice the average.

Orbit and spin. Each planet, including Earth, travels around the sun in a regular orbit—that is, they revolve around the sun in the same direction that the sun rotates, which is counterclockwise as viewed from above Earth's North Pole. Earth orbits the sun at an average distance of about 92,955,808 miles (149,597,870 kilometers). This distance is defined as one astronomical unit (AU), a basic unit that astronomers use in describing the enormous distances in space.

Ancient astronomers thought that the orbits of the planets were circular. It is now known that the orbits are elliptical (elongated), though the orbits of most planets are almost circular. Earth's orbital eccentricity—the extent to which it departs from a perfectly circular path—is very slight. The closest and farthest Earth gets from the sun vary only about 1.7 percent from its average distance from the sun. Among the planets, only Venus and Neptune have orbits that are closer to being perfect circles.

To humans, Earth seems steady and immovable. It gives no sensation of motion, so it is hard to realize how rapidly Earth moves through space in its orbit around the sun. It takes about 365 days, or a whole year, to make one round trip, which seems rather slow. But on average, Earth moves in its orbit at 18.5 miles (29.8 kilometers) per second, or 66,600 miles (107,180 kilometers) per hour. It actually takes about 365.26 days for Earth to travel once around the sun, so 365.26 is the length of an astronomical year on Earth. For convenience, the Gregorian calendar (the calendar in general use) divides most years into 365 days exactly. Every fourth year, with a few exceptions, has 366 days. These longer years are called leap years.

Like the other planets, Earth spins about its axis as well as orbits the sun. Earth makes one rotation on its axis relative to the sun about every 24 hours, 3 minutes, and 57 seconds. In other words, it is about 24 hours from high noon on one day to high noon on the next. So, 24 hours is about the length of a solar day on Earth. Mars has a very similar solar day to Earth's. On the other hand, Jupiter and Saturn rotate more than twice as quickly as Earth and Mars do.

Earth rotates in a counterclockwise direction as viewed from above the North Pole looking down. This is the same direction in which Earth revolves around the sun. Most of the other planets also rotate in this direction, which is called prograde.

Earth's spin axis is not perpendicular, or upright, relative to the plane in which it orbits. Rather, it is tilted about 23.5 degrees. This inclination is mainly responsible for the cycle of seasons on Earth. As Earth orbits the sun, the North Pole always points in the same direction in space. As a result, during some of the year, the North Pole is tilted away from the sun. It is then winter in the Northern Hemisphere and summer in the Southern Hemisphere. Six months later the situation is reversed, and the North Pole is tilted toward the sun and the South Pole away from it.

The familiar, near side of the moon appears in an image taken in 1992 by the Galileo spacecraft while on its way to Jupiter. The lighter areas are heavily cratered, ancient highlands, while the darker areas are younger, lava-filled impact basins.

NASA/JPL/Caltech (NASA photo # PIA00405)

Atmosphere

Earth has a relatively thin atmosphere, which is commonly called the air. It is the envelope of gases that extends from the surface up to interplanetary space. The attraction of the planet's gravity keeps the gases from escaping into space. Near the surface, the atmosphere is about 78 percent nitrogen and 21 percent oxygen by volume. The remaining 1 percent is composed of small amounts of many different gases, including argon, water vapor, carbon dioxide, and methane. The atmosphere also includes tiny solid and liquid particles, which are suspended in the gases.

The atmosphere is thick enough so that animals can breathe easily and plants can take up the carbon dioxide they need. The atmosphere is not so dense, however, that it blocks out too much sunlight. Although clouds often appear in the sky, on average enough sunlight reaches the surface so that plants flourish. Through a process called photosynthesis, green plants (and some other organisms) convert the energy of sunlight into the chemical energy of their own bodies. This interaction between plants and the sun is the basic source of energy for nearly all forms of life. It is also the process through which green plants produce the oxygen that animals need to breathe.

As mentioned above, the high level of oxygen in Earth's atmosphere is unusual. Oxygen is very reactive, which means that it readily combines with other chemical elements. Under most planetary conditions, it would be taken up in the formation of chemical compounds in the atmosphere, surface, and crust. The free oxygen in Earth's atmosphere would not exist without green plants. When supplied with sunlight, the plants take in carbon dioxide and give off oxygen. Animals and some bacteria take in oxygen as part of respiration and give off carbon dioxide.

The weight of the atmosphere as it presses on Earth's surface is great enough to exert an average force of about 14.7 pounds per square inch (1.03 kilograms per square centimeter) at sea level. The pressure is lower at higher altitudes, because there is less atmosphere pressing down from above. The density of the atmosphere also decreases greatly with altitude. Most of its mass is concentrated in its lower and middle levels.

To describe the atmosphere, scientists divide it into five main regions, or layers, based on differences in temperature and other properties. The layers surround Earth like a set of spherical shells. The lowest layer is called the troposphere. It extends from the surface up to about 6–9 miles (10–15 kilometers), depending on the latitude and season. In this layer, temperatures drop rapidly with altitude. Nearly all the planet's water vapor, and thus most clouds, exists in the troposphere. It is also where most weather occurs. Temperature differences make the air in the troposphere relatively unstable. Sunlight heats air near the surface, and the lighter, warmer air tends to rise. Colder air, which is denser, tends to sink. This process is called convection.

In the region directly above the troposphere, the temperature increases slowly with altitude. This layer, called the stratosphere, extends up to about 30 miles (50 kilometers) above Earth's surface. The stratosphere contains a small concentration of ozone. This "ozone layer" shields Earth by absorbing most of the incoming ultraviolet radiation from the sun, which is harmful to living things.

Above the stratosphere is the mesosphere, where the temperature again decreases with height. The top of the mesosphere, which is about 50–55 miles (80–90 kilometers) above the surface, is the coldest part of the atmosphere. Above it is the thermosphere, in which temperatures again get warmer with altitude. The gases are very thin in this layer. The huge uppermost layer, called the exosphere, extends from about 300 miles (500 kilometers) above Earth's surface up to interplanetary space. In the exosphere the density of gases is extremely low. Hydrogen and helium, the lightest gases, predominate.

In the upper regions of the atmosphere, especially in the thermosphere and above, many of the molecules and atoms of Earth's atmosphere are ionized. That is, they carry either a positive or negative electrical charge. This part of the atmosphere is called the ionosphere. It is in the ionosphere that auroras occur.

Luminous layers of the atmosphere appear above the dark edge of Earth, at bottom, in an image captured at sunset by astronauts aboard the International Space Station. The orange and red layer is the lowest and densest level of the atmosphere, called the troposphere, while the blue layer is the stratosphere. Ozone in the stratosphere shields Earth from harmful ultraviolet radiation from the sun.

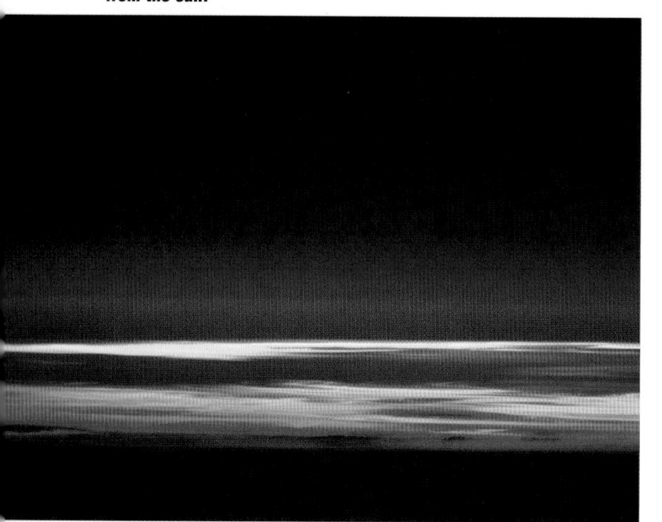

Courtesy, Image Science & Analysis Laboratory, NASA Johnson Space Center, No. ISS001-421-24

Hydrosphere

The huge volume of water at or near Earth's surface can be considered a discontinuous layer called the hydrosphere. Like the atmosphere, it is essential for life on Earth. The hydrosphere includes all the liquid and frozen water on or near the surface. This "layer" also includes the gaseous water vapor in the air and the groundwater, or the water held below the surface in rocks and soil. The hydrosphere extends from about 3 miles (5 kilometers) below the ground to about 9 miles (15 kilometers) up in the air. The largest component of the hydrosphere is by far the oceans. Salty oceans cover some 71 percent of Earth's surface and have a volume of about 336 million cubic miles (1.4 billion cubic kilometers). They account for more than 97 percent of the hydrosphere's volume.

Energy from the sun causes water to continually cycle between Earth and its atmosphere. This circulation is known as the hydrologic cycle. It involves the transfer of water from the oceans through the air to the continents and back to the oceans again. The water is transferred both over and beneath the land surfaces. For example, through a process called evaporation, the sun's heat causes a small percentage of water from the surface of the oceans to become water vapor in the air. Atmospheric circulation carries some of that vapor over land, where some of the vapor condenses into clouds and falls to the surface as rain or snow. Some of the water seeps underground, while some may enter lakes or rivers, which carry water back to the oceans. In reality, however, the hydrologic cycle is much more complex and involves many more processes than this example shows.

Surface and Interior

More than 90 percent of Earth's mass consists of iron, oxygen, silicon, and magnesium, elements that can form the minerals known as silicates. These elements are not evenly distributed throughout the planet but are found in varying concentrations. Earth's consistency varies largely according to distance from the center. These differences are partly the result of differences in pressure and temperature, which increase greatly with depth. Also, as the planet formed some 4.6 billion years ago, it partially melted, which caused different materials to separate. Through a process called differentiation, denser materials such as iron mostly sank to the center. Lighter materials, such as silicates, rose to the outside.

Earth's structure comprises three basic layers. The outermost layer, which covers the planet like a thin skin, is called the crust. It includes the planet's rocky surface. Beneath that is a thick rocky layer called the mantle. The innermost region is the metallic core. Each layer is subdivided into other, more complex, structures.

Much scientific study has been devoted to the thin crustal area on which humans live, and most of its surface features are well known. However, scientists cannot rely on direct observation when studying the planet's interior. Much of the information about Earth's inner layers comes from the study of seismic waves, which are vibrations generated during earthquakes and underground explosions. These waves move through different types of material at different speeds. Also, abrupt differences in density at the boundaries between different layers cause the waves to change direction. By analyzing the velocity

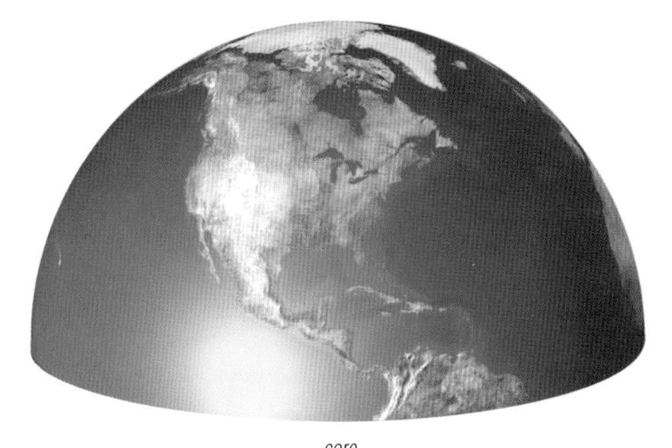

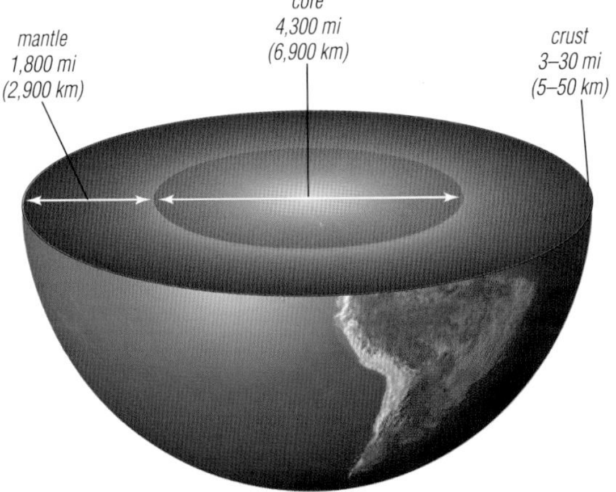

Earth's interior has three main layers: a metallic core; a thick, rocky mantle; and a thin, rocky crust. The measurements provided are broadly rounded averages.

and other properties of seismic waves, scientists develop models of the planet's interior structure.

Crust. The planet's crust is composed mainly of silicate rocks. It varies in thickness from place to place, but it is thin everywhere. The crust accounts for less than half of 1 percent of Earth's mass. There are two basic types of crust: oceanic crust and continental crust. The average thickness of the crust under the oceans is only about 3 to 6 miles (5 to 10 kilometers). The crust that makes up the continents is thicker, extending some 30 miles (50 kilometers) below the surface.

These two kinds of crust also consist of different kinds of rocks. The oceanic crust is denser, being formed mainly of basalt. The lower-density continental crust consists mostly of granite and similar rocks. The rocks of both types of crust actually float on top of the denser rocks of the mantle.

Mantle. The mantle extends roughly to a depth of about 1,800 miles (2,900 kilometers). It accounts for most of Earth's mass. The uppermost part of the mantle is solid. Together, the crust and the very top of the mantle form the planet's rigid outer shell, which is called the lithosphere.

A large part of Earth's interior consists of the lower mantle. Temperatures and pressures are much greater there than nearer the surface. Over the vast periods of

Geometry and motions of the Earth-moon system

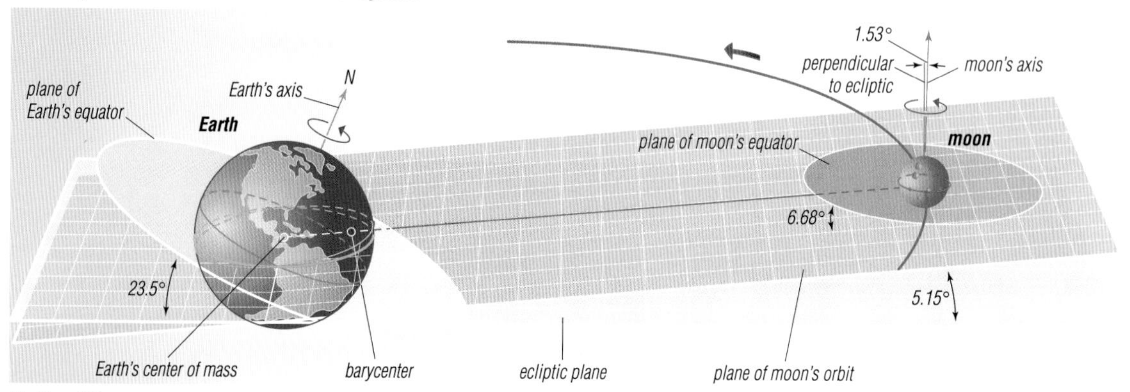

Earth and its moon exert a strong gravitational pull on each other, forming a system with complex properties and motions. Although it is commonly said that the moon orbits Earth, the two bodies actually orbit each other about a common center of mass, called the barycenter. The barycenter lies within the outer portion of Earth's interior. As the Earth-moon system orbits the sun, both bodies also rotate on their axes. Earth's axis is tilted about 23.5° from the ecliptic plane, which is the imaginary plane in which its orbit lies.

geologic timescales, it behaves as a fluid and responds to stress by very slowly flowing.

Core. About a third of Earth's mass is in its dense core. The core extends outward from the center to a radius of nearly 2,200 miles (3,500 kilometers). It is probably composed mostly of iron, with some nickel and other elements. The outer core is liquid, while the smaller inner core is solid. Seismic evidence suggests that the inner core rotates a bit faster than the rest of the planet. Temperatures at the very center of Earth are thought to reach some 8,500 to 12,100° F (4,700 to 6,700° C), which is about as hot as the surface of the sun.

Plate tectonics. Earth's lithosphere—the crust and the uppermost part of the mantle—is broken into many rigid blocks called plates. There are about 12 major plates and several smaller ones. Like the crust, the plates are of two types: continental and oceanic. Both types of plate are constantly in motion, mainly horizontally, relative to one another. They slide over a layer of the mantle that is thought to be partially molten. The plates interact with one another at their boundaries, either spreading apart, coming together, or sliding past one another. This process is known as plate tectonics.

Although the motion of the plates is relatively slow—typically about only 2 to 4 inches (5 to 10 centimeters) per year—it has dramatic and often violent effects. Plate tectonics is responsible for most of the planet's earthquakes and volcanic eruptions, which occur mainly along plate boundaries. In addition, when continental plates push against each other, they can pile up the land to form high mountains, such as the Himalayas. When plates pull away from one another, they can open huge rifts in the land or cause oceans to form. Plate tectonics also causes continental drift, as the moving plates carry the continents into new configurations.

Most scientists believe that the motion of the plates is caused in some way by a process of heat transfer in the mantle called convection. The decay of radioactive elements is thought to heat some of the material in the mantle. Currents of the hotter, more buoyant material rise, while currents of cooler, denser materials sink. Similar movements can be seen in a pot of boiling water, as well as in the lower layer of Earth's atmosphere.

Magnetic Field and Magnetosphere

Of the solar system's four inner planets, only Mercury and Earth currently have global magnetic fields. Earth's field is about a hundred times stronger than Mercury's. Scientists believe that planetary magnetic fields form from electrical currents generated by fluid motions in a planet's core. Earth's large, hot, rapidly spinning core is probably responsible for its strong magnetic field.

Near the surface, Earth's magnetic field has two poles, north and south, as if the planet were a huge bar magnet. Magnetic compasses used to find directions work because of this magnetic field. Periodically, the two poles switch polarities, so that the north magnetic pole becomes the south magnetic pole and vice versa. This seems to happen at intervals ranging from tens of thousands of years to millions of years.

The magnetic field dominates a large teardrop-shaped area around Earth. The region of space dominated by a planet's magnetic field is known as its magnetosphere. Earth's magnetosphere is shaped by interactions with the solar wind, or streams of electrically charged particles coming from the sun. The magnetic field holds off the solar wind about 40,000 miles (65,000 kilometers) above the planet. This collision creates a shockwave that deflects the solar particles around and past Earth. Earth's magnetic field is slightly squeezed in on the side that faces the sun

From Robin M. Canup, "Simulation of a Late Lunar-Forming Impact," *Icarus*, vol. 168 (2004)

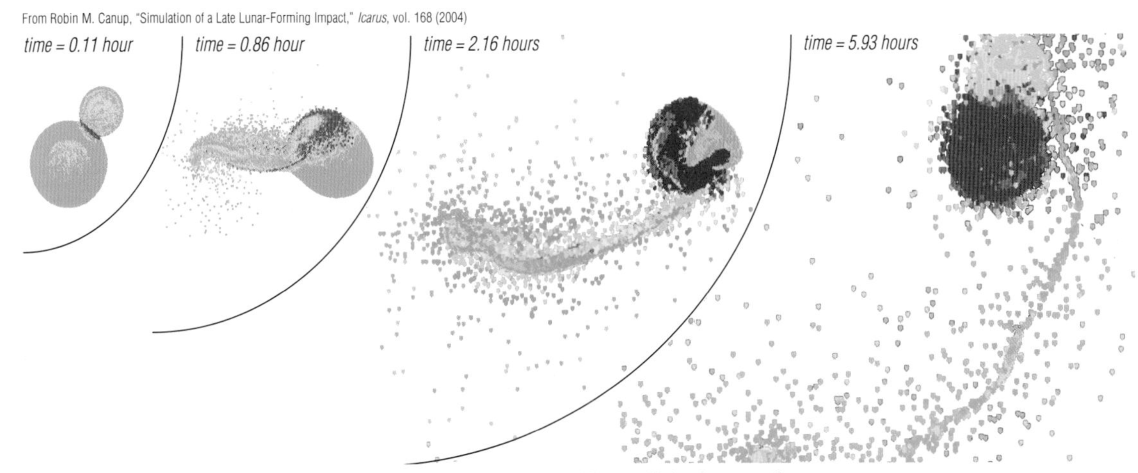

A computer simulation shows how the moon may have formed from a violent collision between the forming Earth and a smaller object about the size of Mars. Four stages in the first six hours of the event are shown. The material scattered by the collision is thought to have come mainly from the mantles of both bodies. This scattered material eventually clumped together to form the moon. The different colors in the simulation show relative temperatures of the material heated by the collision.

and is pulled out into a long tail on the side away from the sun. The tail of the magnetosphere extends millions of miles away from the sun.

Earth's magnetic field traps highly energetic electrically charged particles—electrons and protons—in two large concentric clouds above the planet. These orbiting swarms of particles are called the Van Allen radiation belts. The charged particles are very thinly distributed in these regions. Their discovery in 1958 via Earth-orbiting satellites was one of the earliest accomplishments of the space age.

The Van Allen "belts" are actually doughnut-shaped regions. The particles within them travel in a complex corkscrew pattern while they also move between the Northern and Southern hemispheres. At the same time, each whole group slowly drifts either westward or eastward around Earth. Some of the particles in the Van Allen radiation belts come from the solar wind. Other particles in the belts are thought to be produced when Earth's atmosphere is struck by cosmic rays, which are highly energetic charged particles mainly from beyond the solar system. The atmosphere shields the planet from the radiation in these regions.

The Moon

Earth's moon is a nearly spherical, rocky body with a diameter of about 2,160 miles (3,476 kilometers). It is about a quarter the size of Earth and about 80 times less massive. Nevertheless, it is the largest and most massive moon in the solar system relative to its planet. Scientists believe that the moon is so large compared to Earth because it formed in a different way than most other moons. According to this theory, in the early era of the solar system, a rocky body about the size of Mars crashed into the newly forming Earth. The collision knocked off chunks of matter from the outer layers of both bodies. These fragments were captured by Earth's gravity and began orbiting the planet. The fragments clumped together into a larger body, which swept up the smaller debris in its path. This object ultimately developed into the moon.

The surface gravity of the moon is about a sixth of that of Earth. Because the moon is so close to Earth, the pull of its gravity causes the tides. The tides are most obvious as the regular, periodic rise and fall of the surface of Earth's seas. The tides also regularly cause very slight deformations in the solid parts of the planet. Earth's much stronger gravity affects the moon as well. Its pull causes slight bulges on the moon's surface in the direction of Earth. In addition, gravitational effects of the Earth-moon system are very slowly causing Earth to spin slower and the moon to recede from Earth.

The moon orbits Earth at an average distance of about 238,900 miles (384,400 kilometers). It completes one trip around Earth about every 27.3 days. Like most other large moons, Earth's moon takes the same amount of time to complete one rotation about its axis as it does to complete one orbit. Because of this, the moon always shows nearly the same side to Earth. The far side of the moon was first seen in photographs taken in 1959 by the Soviet space probe Luna 3. The first humans to see the lunar far side directly were the astronauts of the United States Apollo 8 mission, which circled the moon three times in 1968.

The sun always illuminates half of the moon, the half that is pointed toward the sun at that moment. However, the varying positions of Earth and the moon as they orbit the sun cause different amounts of the sunlit part of the moon to be visible from Earth. This causes the moon's changing appearance from Earth, called its phases. The phases occur in a cycle. At the beginning of the cycle, the moon cannot be seen from Earth. This phase is called the new moon. Later, the moon appears as a thin crescent and then a half disk. It appears as a full disk, called the full moon, when

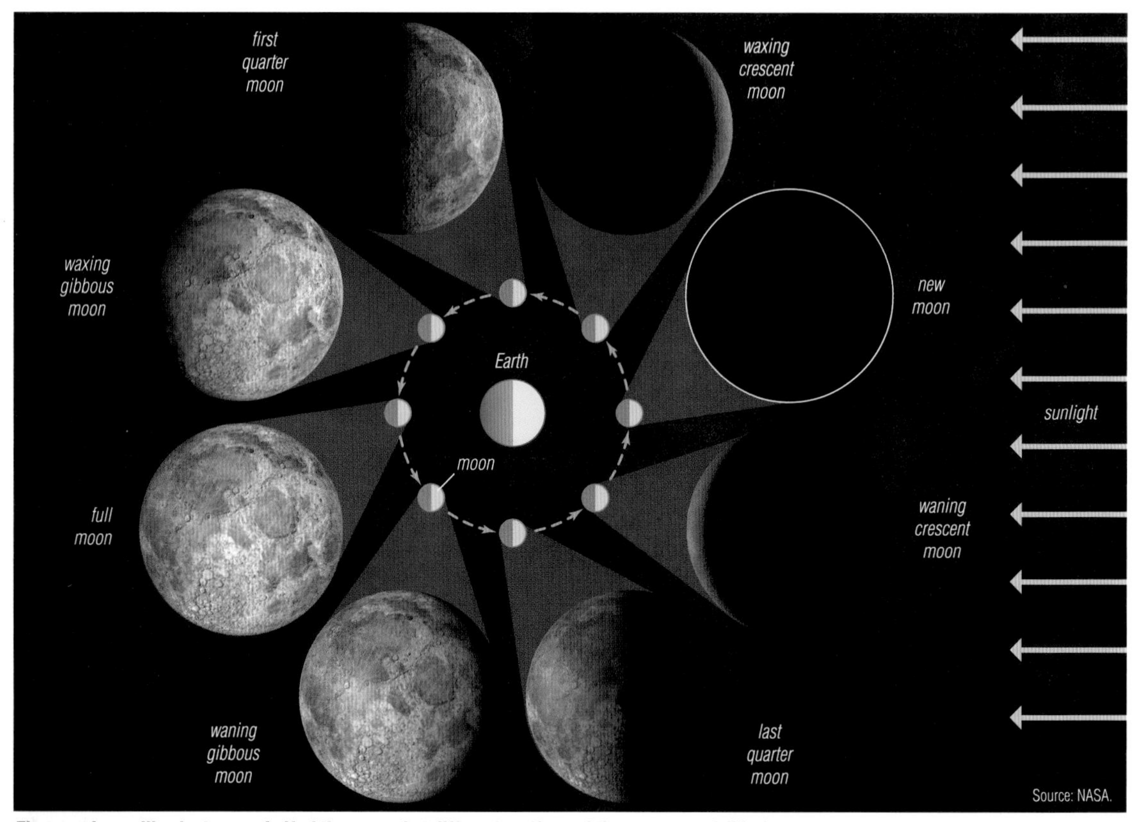

The sun always illuminates one half of the moon, but different portions of the moon are visible from Earth as the moon orbits Earth and the two bodies circle the sun. A diagram shows the position of the moon in each of its phases, relative to Earth and the sun. The photographs show how the moon appears to observers on Earth during each phase.

the entire sunlit half of the moon can be seen from Earth. Over the following days, a smaller and smaller portion appears. Each cycle of phases, from new to full and back to new again, lasts about 29.5 days.

The moon is only about two thirds as dense as Earth. It is composed mainly of silicate rocks, and it probably has a small metallic core. Its surface is pocked with many craters and is strewn with dust and rocky rubble. This is the result of numerous collisions with asteroids and comets, as well as smaller chunks of matter. In ancient times large crashes formed enormous impact basins on the moon. The basins were filled in with lava a few billion years ago. Since then, the moon has cooled off, and it no longer experiences volcanic activity. The lava-filled basins are called maria (with 'mare' being the singular form). They are the dark areas of the moon that one can see from Earth. The lighter areas are highlands.

As Earth's nearest neighbor, the moon has fascinated people since ancient times. In the 1960s and early 1970s, it became the first alien world that humans visited. In all, 12 humans have walked on the moon, and it has been the target of several dozen spacecraft.

Facts About the Moon

Average Distance from Earth. 238,900 miles (384,400 kilometers).

Diameter at Equator. 2,160 miles (3,476 kilometers).

Period of Orbit Around Earth.* 27.3 Earth days.

Rotation Period.† Same as orbital period.

General Composition. Silicates.

Average Surface Temperature. Day: 224° F (107° C); Night: –244° F (–153° C).

Weight of Person Who Weighs 100 Pounds (45 Kilograms) on Earth. 16.5 pounds (7.5 kilograms).

*Sidereal revolution period, or the time it takes the moon to revolve around Earth once, relative to the fixed (distant) stars.

†Sidereal rotation period, or the time it takes the moon to rotate about its axis once, relative to the fixed (distant) stars.

Topographic maps (on opposite page) of the moon's near side (top) and far side (bottom) show variations in elevation across its terrain. The lowest areas, which have been colored purple, are about 9.9 kilometers (6 miles) below the moon's average radius (like sea level on Earth). The highest areas, colored red, are some 8 kilometers (5 miles) above the moon's average radius. Labels indicate selected major topographic features and the landing sites of the Apollo spacecraft. The maps were made with data and images collected by the Clementine spacecraft as it orbited the moon in 1994.

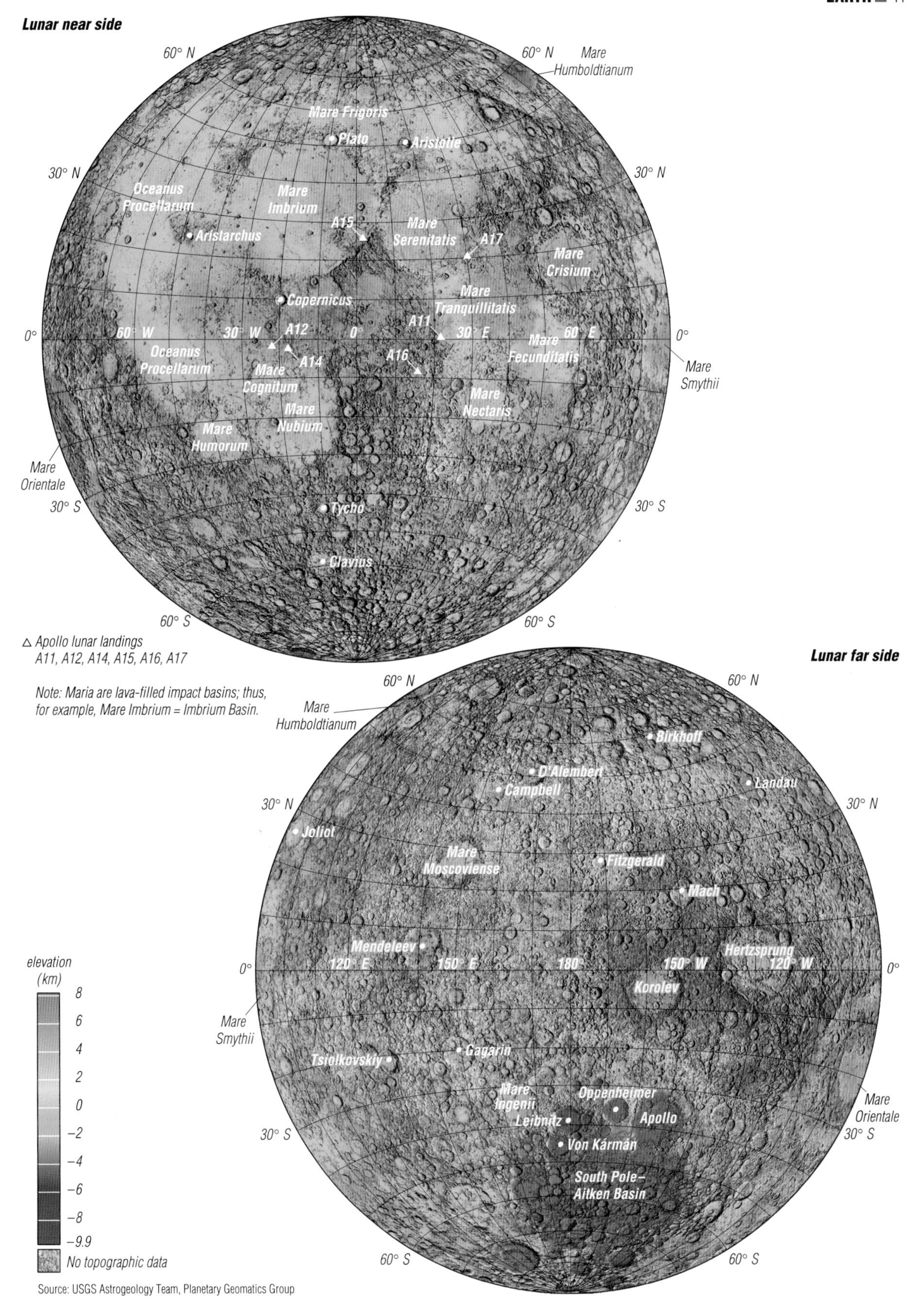
Lunar near side
60° N
60° N
Mare Humboldtianum
Mare Frigoris
Plato
Aristotle
30° N
30° N
Oceanus Procellarum
Mare Imbrium
A15
Mare Serenitatis
A17
Aristarchus
Mare Crisium
Copernicus
Mare Tranquillitatis
0°
60° W
30° W
A12
0°
A11
30° E
60° E
0°
Oceanus Procellarum
A14
A16
Mare Fecunditatis
Mare Smythii
Mare Cognitum
Mare Nectaris
Mare Nubium
Mare Humorum
Mare Orientale
30° S
30° S
Tycho
Clavius
60° S
60° S
△ Apollo lunar landings
A11, A12, A14, A15, A16, A17
Note: Maria are lava-filled impact basins; thus, for example, Mare Imbrium = Imbrium Basin.
Lunar far side
60° N
60° N
Mare Humboldtianum
Birkhoff
D'Alembert
Campbell
Landau
30° N
30° N
Joliot
Mare Moscoviense
Fitzgerald
Mach
Mendeleev
Hertzsprung
0°
120° E
150° E
180°
150° W
120° W
0°
Korolev
Mare Smythii
Tsiolkovskiy
Gagarin
Mare Ingenii
Oppenheimer
Mare Orientale
Leibnitz
Apollo
30° S
30° S
Von Kármán
South Pole–Aitken Basin
60° S
60° S
elevation (km)
8
6
4
2
0
−2
−4
−6
−8
−9.9
No topographic data
Source: USGS Astrogeology Team, Planetary Geomatics Group

MARS

The fourth planet from the sun is Mars. Easily visible from Earth with the naked eye, it has intrigued stargazers since ancient times. It often appears quite bright and reddish in the night sky. Babylonians mentioned Mars in records from about 3,000 years ago, associating the blood-red planet with their god of death and disease. The name Mars is that of the ancient Roman god of war.

Mars is also a nearly ideal subject for observing with a telescope from Earth. Venus approaches more closely to Earth in its orbit. However, Mars also passes relatively near Earth and, unlike Venus, its surface is generally not obscured by thick clouds. Mars is farther from the sun than Earth is, so the planet often appears high in the sky. Venus, on the other hand, can never be seen far from the glare of the sun.

Over the centuries, observers have noted various phenomena on the Martian surface, some of which they thought might be signs of life. For instance, dark markings cover about a third of the surface and change in a seasonal pattern in both extent and color. They were once thought to be vast seas or areas of vegetation. A few astronomers even thought they saw straight lines that could have been canals.

The explanation for these and many other observations had to await the first exploratory space missions in the 1960s and '70s. Meanwhile, the so-called red planet captured the popular imagination as a possible home of alien life—the armies of "little green men" of science fiction stories, movies, and radio and television programs. Scientists now know that there are no manufactured canals on Mars. The changes in the dark areas result largely from dust, which shifts along with the winds. There are no humanoids on the planet, nor even any animals or plants. No forms of life have yet been found on Mars.

Nevertheless, the search for life on Mars continues, in part because it shows signs of having been wetter in the past. Water is necessary for all known forms of life. It is possible that microscopic life once existed on

NASA/JPL/Malin Space Science Systems

Mars, the most Earth-like of the planets, appears in a computer-generated image based on photographs taken by Mars Global Surveyor on a day during the northern hemisphere's summer. At the top of the globe is the northern polar ice cap, while the huge gash just below the equator is the canyon system Valles Marineris. White clouds of water ice surround the peaks of the most prominent volcanoes.

Mars. Today the surface is too cold and the air is too thin for liquid water to exist there for long. It is also bombarded with ultraviolet radiation from the sun, which is very harmful to living things. (Earth's denser atmosphere protects it from most of this radiation.) If any life-form exists on Mars today, many scientists believe it would be tiny organisms in protected niches just below the surface. Whether or not Mars has ever had life, the planet remains an intriguing object of study. Many unmanned spacecraft have achieved outstanding success in obtaining information about this alien world that is in many ways quite Earth-like.

Basic Planetary Data

The orbit of Mars lies between Earth's orbit and the main asteroid belt. Jupiter, which lies beyond the belt, is its very distant outer planetary neighbor. Mars is the outermost of the four inner planets of the solar system, which are Mercury, Venus, Earth, and Mars. The inner planets are also called the terrestrial, or Earth-like, planets. All four are dense, rocky worlds with solid surfaces and few or no moons. Mars has two small moons, Phobos and Deimos.

Size, mass, and density. Mars is the second smallest planet in the solar system, after Mercury. Its diameter at the equator is 4,221 miles (6,792 kilometers), which is only slightly more than half the size of Earth's. Mars's lower density makes the planet only about a tenth as massive as Earth. In fact, Mars's density is closer to that of Earth's moon than to that of the three other inner planets.

Orbit and spin. Like all the planets, Mars travels around the sun in an elliptical, or oval-shaped, orbit. Its orbit is on average about 1.5 times as far from the sun as Earth's. Mars's orbit is more eccentric, or elongated, than Earth's, however, so its distance from the sun varies more. Mars is about 128 million miles (207 million kilometers) from the sun at its perihelion, or the closest point in its orbit to the sun. At its aphelion, its farthest point from the sun, it is some 155 million miles (249 million kilometers) away.

The planet completes one revolution around the sun in about 687 Earth days. In other words, a year on Mars is about 687 Earth days long. That is almost twice the time it takes Earth to complete its orbit of about 365 days.

Mars's distance from Earth varies considerably, from less than 35 million miles (56 million kilometers) to nearly 250 million miles (400 million kilometers). The best time to view Mars from Earth is when it is closest to both the sun and Earth so that it appears both bright and large. The planet is easiest to observe when it is at opposition, or when it is on the opposite side of Earth from the sun. Mars then appears high in Earth's sky, and its full face is lighted. Oppositions of Mars occur about every 26 months. About every 15 years Mars is both close to Earth and in opposition, an arrangement that provides optimal viewing conditions.

Mars rotates on its axis at roughly the same rate as Earth. A Martian day, called a sol, lasts about 24 hours and 37 minutes, which is just a bit longer than an Earth day. Its spin axis is also tilted at an angle similar to Earth's—about 24.9 degrees compared with 23.5 for Earth. As a result, Mars has distinct seasons like Earth does. First one hemisphere, then the other receives more sunlight during the planet's orbit around the sun. Because Mars has a more eccentric orbit than Earth's, Martian seasons are less equal in length. The Martian summer, for example, is about 16 percent longer in the north than in the south. (*See also* Planet, "Years, days, and seasons.")

Facts About Mars

Average Distance from Sun. 141,636,000 miles (227,941,000 kilometers).

Diameter at Equator. 4,221 miles (6,792 kilometers).

Average Orbital Velocity. 15.0 miles/second (24.1 kilometers/second).

Year on Mars.* 687.0 Earth days.

Rotation Period.† 24.6 Earth hours.

Day on Mars (Solar Day). 24.7 Earth hours.

Tilt (Inclination of Equator Relative to Orbital Plane). 24.9°.

Atmospheric Composition. 95.3% carbon dioxide; 2.7% molecular nitrogen; 1.6% argon; 0.13% molecular oxygen, 0.07% carbon monoxide; 0.03% water vapor; trace amounts of neon, krypton, xenon, other gases.

General Composition. Iron and sulfur in core, silicate rocks.

Average Surface Temperature. –82° F (–63° C).

Weight of Person Who Weighs 100 pounds (45 kilograms) on Earth. 38.0 pounds (17.2 kilograms).

Number of Known Moons. 2.

*Sidereal revolution period, or the time it takes the planet to revolve around the sun once, relative to the fixed (distant) stars.

†Sidereal rotation period, or the time it takes the planet to rotate about its axis once, relative to the fixed (distant) stars.

Telescopic observers have noted distinctive bright and dark features on Mars for hundreds of years. One of the sharpest images taken from the vicinity of Earth, photographed by the Hubble Space Telescope, captures the red planet on the last day of spring in the northern hemisphere. The large, dark marking just below and to the east of center is Syrtis Major. Beneath it is the giant impact basin Hellas, covered with an oval of white clouds.

NASA/JPL/David Crisp and the WFPC2 Science Team

Atmosphere, Surface, and Interior

The planet's atmosphere, or surrounding layers of gases, is not very dense. Its surface is a cold, dusty desert. It is possible that water might periodically seep up in places from ice beneath the surface. Under such conditions, however, liquid water on the surface would very quickly freeze or vaporize. But evidence suggests that in the remote past the planet may have been more Earth-like, with a thicker atmosphere and warmer surface temperatures. Large bodies of water may have been found on the Martian surface for a while in its early history. And some evidence suggests that at least some water flowed on the surface in the geologically recent past.

Atmosphere. The Martian atmosphere is very thin. It exerts less than a hundredth the surface pressure of Earth's atmosphere. Like Venus' atmosphere, it consists almost entirely of carbon dioxide. Venus' atmosphere is extremely dense, however, so the total volume of carbon dioxide is much greater there. Near the surface of Mars, carbon dioxide accounts for about 95.3 percent of the atmosphere by weight. Most of the rest of the atmosphere consists of nitrogen and argon, with small amounts of oxygen, carbon monoxide, water vapor, neon, krypton, xenon, and other gases. Trace amounts of methane also have been detected. A large amount of dust particles is suspended in the gases.

The lower Martian atmosphere is roughly as cold as the air on Earth above Antarctica in the winter. It is typically about –100° F (–70° C). Clouds, haze, and fog are common in the air near the ground, especially over valleys, craters, volcanoes, and other areas of low or high surface elevation. Most of these low-lying clouds are formed of water ice. In addition, very thin clouds, perhaps of carbon dioxide, are found quite high up in the atmosphere.

At the surface the winds are generally light. The average wind speeds are typically less than 4.5 miles (7.2 kilometers) per hour, though gusts sometimes blow as fast as 90 miles (144 kilometers) per hour. In winter Mars has strong jet streams like Earth's, which blow westward high in the atmosphere. These winds are much lighter during the Martian spring and summer. Unlike on Earth, the air circulation on Mars also has a fairly strong north-south pattern, which transfers air from pole to pole.

Rapidly swirling columns of dust, called dust devils, have been seen whirling along the surface. Dust storms also occur frequently on the planet. They are especially common in the southern hemisphere in spring and summer, when the surface is warmest. About every two or three years, Mars is engulfed by global dust storms. Local temperature differences generate strong winds that lift dust from the surface. The thick dust clouds block the sunlight, gradually causing the surface temperatures to even out and the winds to subside. Some of the atmospheric dust is deposited in a "snowfall" of dust and ice in the polar regions.

Ice caps are found at both the north and south poles. The caps alternately grow and shrink according to seasonal changes. Each ice cap grows larger when its hemisphere experiences fall and shrinks during its spring. Most of the ice then "burns off," escaping into the air as a gas. As a result, the atmospheric pressure increases above the shrinking cap. As summer approaches, the ice cap shrinks into several small patches of ice. In fall, as gases from the air again condense and freeze to form the larger cap, the atmospheric pressure decreases over the cap.

A whirlwind of dust, called a dust devil, appears in an image captured by Mars Global Surveyor. Its camera was pointed essentially straight down above the Martian surface. As the dust devil traveled, it left a faint track, at left, and cast a long shadow, at right. The dust cloud itself appears as a foreshortened white column, at center.

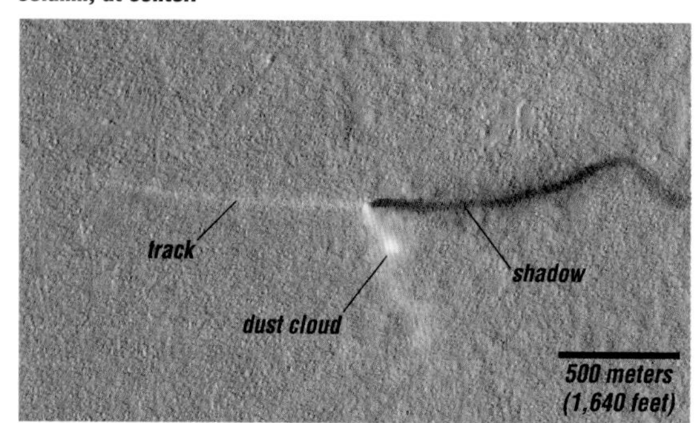

NASA/JPL/Malin Space Science Systems

The permanent cap of water ice at the Martian north pole (top) appears in an image taken by Mars Global Surveyor during early summer in the north. Just south of the north pole, at about 70.5° N., the Mars Express spacecraft found a patch of water ice sheltered on the floor of a crater (bottom) during late summer. The crater is about 22 miles (35 kilometers) across and at its deepest extends about a mile (2 kilometers) beneath the rim. The image has been processed by computer, however, to provide a perspective view that exaggerates the vertical relief by three times.

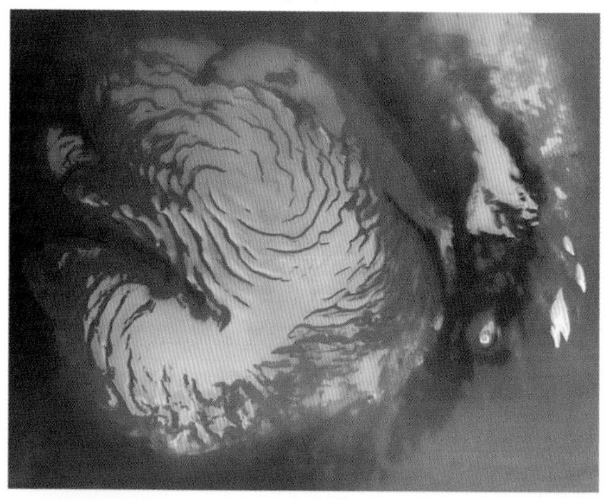

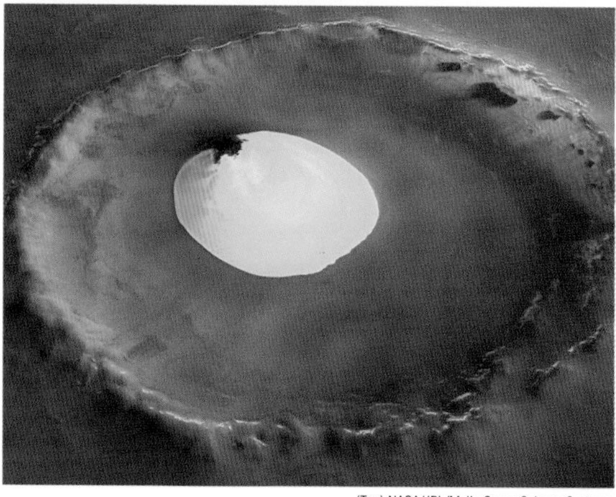

(Top) NASA/JPL/Malin Space Science Systems, (bottom) ESA/DLR/FU Berlin

The two caps differ somewhat in size and composition. The southern cap is larger. It extends to about 50° S. at its greatest extent, compared with 55° N. for the northern ice cap. The north pole has a small permanent cap of water ice. It temporarily accumulates larger areas of carbon dioxide ice (or "dry ice") in fall and winter. The south pole's cap seems to contain some carbon dioxide ice and some water ice year-round (which grow and shrink seasonally).

Surface. The thin Martian atmosphere does not shield the surface from ultraviolet radiation from the sun. It also does not insulate the planet as well as Earth's thicker one does. The surface of Mars is thus colder than Earth's would be if the two planets were the same distance from the sun. The temperature of Martian air at about human height varies widely over the course of a day, from about −119° F (−84° C) before dawn to about −28° F (−33° C) in the afternoon. The temperature of the surface itself is colder.

The rocks and soil on the Martian surface are typically rusty reddish brown because they contain much iron oxide. The planet's surface features include smooth, desolate plains and windblown sand dunes, deep, rugged canyons and steep cliffs, rolling hills, mesas, and enormous volcanoes. Several areas have winding channels that were probably carved by ancient floodwaters. The planet conspicuously lacks features such as long mountain chains that on Earth form by plate tectonics (*see* Earth, "Plate tectonics"). Instead, many Martian features seem to have been formed by fracturing of the planet's crust, volcanic

Photo NASA/JPL/Caltech (NASA photo # PIA00300)

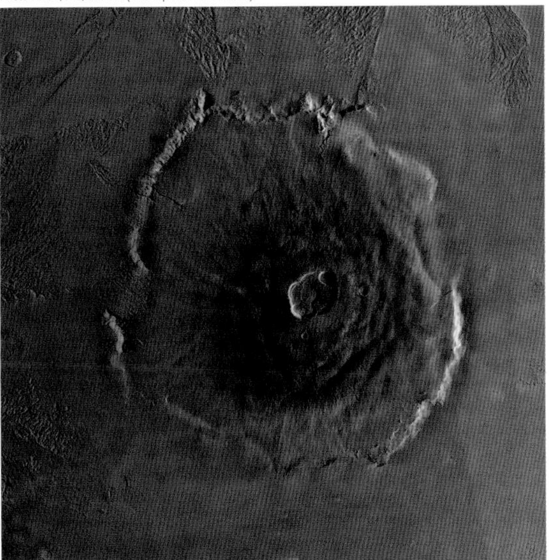

The Martian volcano Olympus Mons is the largest known volcano in the solar system. It appears in a mosaic of images taken from above by the Viking 1 orbiter.

A map of the topography of Mars shows the contrast between the heavily cratered highlands in the southern hemisphere and the smoother lowlands in the northern hemisphere. The map is a Mercator projection made with data collected by Mars Global Surveyor. The topographic relief has been color coded according to the key at right. The lowest elevations, which are colored purple, lie some 8 kilometers (5 miles) below the average reference level (similar to sea level on Earth). The highest elevations, colored white, are more than 12 kilometers (7.5 miles) above the average value. Labels identify selected major features and the landing sites of space probes.

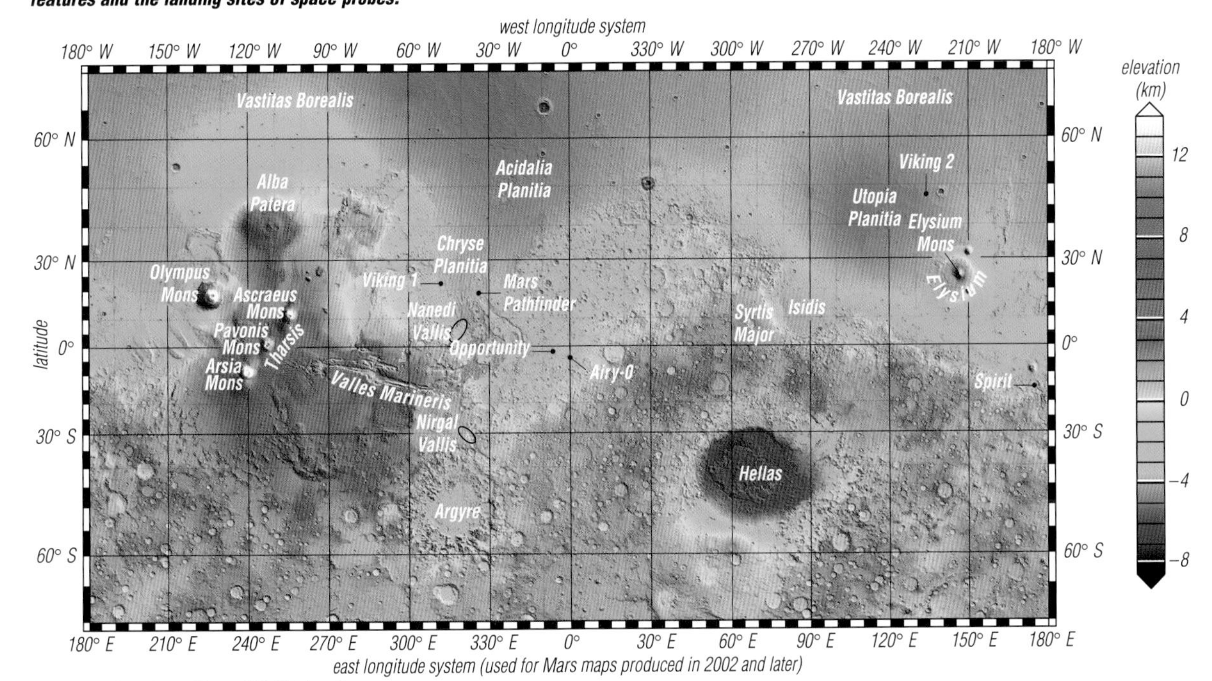

Source: Mars Orbital Laser Altimeter (MOLA) Science Team

(Top) Photo NASA/JPL/Caltech (NASA photo # PIA00422); (bottom) NASA/JPL/Malin Space Science Systems

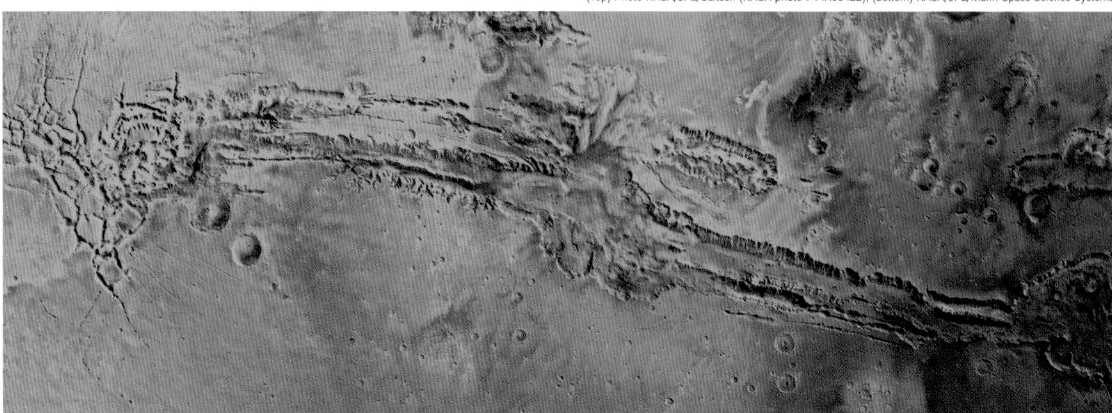

activity, or the wearing away or depositing of material by the wind.

Even though Mars is much smaller than Earth, it has a greater variation in surface elevation. The lowest point on Mars is Hellas, a giant impact basin. It is probably the result of an asteroid colliding with the planet early in Martian history. The planet's highest point is the volcano Olympus Mons. The difference in elevation between these two features is about 18 miles (29 kilometers). By comparison, on Earth the difference in elevation between the lowest point (the Mariana Trench) and the highest point (Mount Everest) is some 12.4 miles (20 kilometers).

The planet's northern and southern hemispheres are strikingly different, for reasons that are not yet clear. Much of the south consists of heavily cratered highlands, while flat lowlands with few craters predominate in the north. Craters and impact basins form when asteroids or other chunks of matter crash into a planet's surface. Planetary scientists can estimate the age of a planet's surface by counting the number of these scars. The older the surface, the more time it has potentially been exposed to falling objects and debris. In other words, surfaces with many craters are generally older than those with few craters. Earth's moon, which is also heavily cratered, was most intensely bombarded with asteroids and large fragments before about 3.8 billion years ago. The southern surface of Mars is probably similarly ancient.

The northern lowlands of Mars are younger. They include the broad plains called Chyrse Planitia, Acidalia Planitia, and Utopia Planitia. Some scientists think that the smooth northern plains may once have been the beds of seas that were filled by large floods, but this theory remains controversial.

Some of the red planet's most prominent features lie along the borders between the northern and southern hemispheres. Among them is Tharsis, a broad, high volcanic dome covered with lava flows. Three of the planet's largest volcanoes are at the top of the dome, while Olympus Mons is just to the northwest. The largest known volcano in the solar system, Olympus Mons reaches a height of 13 miles (21 kilometers) above the average reference altitude (like sea level on Earth). This makes the volcano more than twice as high as Earth's Mount Everest. The volcano is very broad, stretching across about 335 miles (540 kilometers). Scientists believe that Mars is almost certainly still volcanically active, though at very low levels.

Valles Marineris is the largest canyon system on Mars. A composite of images (top) taken by the Viking 1 and 2 orbiters shows the full extent of the system, which is some 2,500 miles (4,000 kilometers) long. Closer-up, an image taken by Mars Global Surveyor (bottom) shows outcroppings of sedimentary rock layers in one of the canyons. Some scientists have interpreted these formations as evidence that lakes once partially filled the canyons.

The large bulge of Tharsis has stressed and cracked the surface nearby. An extensive system of fractures surrounds the dome. The largest fracture system is Valles Marineris, an enormous series of connected canyons east of Tharsis. The canyon system is about 2,500 miles (4,000 kilometers) long, which is about 20 percent of the planet's circumference. Its central depression reaches a depth of about 5.6 miles (9 kilometers), which is more than five times deeper than Earth's Grand Canyon. Sediments piled up in the canyons suggest they may once have been filled with lakes.

Water. The Martian surface today is dry and dusty. The question of whether and when liquid water has flowed on the surface of Mars is of particular interest to scientists trying to determine if life has ever existed on the planet. Liquid water is a requirement for all known forms of life. (But, of course, water does not in itself indicate the presence of living things.)

Water currently exists on Mars as small amounts of vapor in the atmosphere, as ice at the poles, and as ice in large regions just below the surface. Rivers, lakes, and even seas may have been present on Mars in its remote past, when it was probably warmer and the atmospheric pressure higher. Several features on the surface appear to have been formed by water, either fed by rainfall or groundwater. Winding valleys on Mars look like river beds and other drainage systems on Earth. Some Martian channels appear to have been completely filled with ancient floodwaters. In addition, some Martian rocks have mineral compositions that suggest they formed through interaction with liquid water.

As mentioned above, Mars's current cold temperatures and low air pressure mean that today liquid water would not last long at the surface. However, photographs taken in the 2000s by the orbiting spacecraft Mars Global Surveyor showed hundreds of

A Mars Express image (top left) reveals what might be a fractured sea of frozen water lying just below the surface at about 5° N. of the equator in Elysium Planitia. The irregular platelike shapes resemble rafts of pack ice found in Earth's polar regions. The view was generated by a computer and based on photographs taken by Mars Express. Many surface features on Mars seem to have been formed by liquid water. Three outflow channels (top right) appear in a view taken by Mars Global Surveyor near the eastern edge of the giant impact basin Hellas. Scientists believe that the channels were carved long ago by floodwaters moving downslope, toward the bottom of the image, into Hellas. Gullies line the steep wall of a crater (bottom left) in Newton Basin in Sirenum Terra in an image taken by Mars Global Surveyor. The gullies in the image and others like it are surprising, because they may have been formed by groundwater seeping up in geologically recent times (though the gullies' source remains controversial). The steep-walled channel Nanedi Valles (bottom right), photographed by Mars Express, may have formed from flowing water. It is one of many Martian valley networks that resemble river valley systems on Earth.

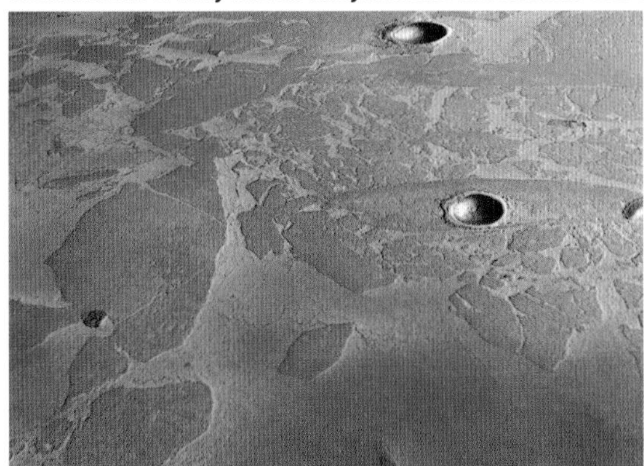

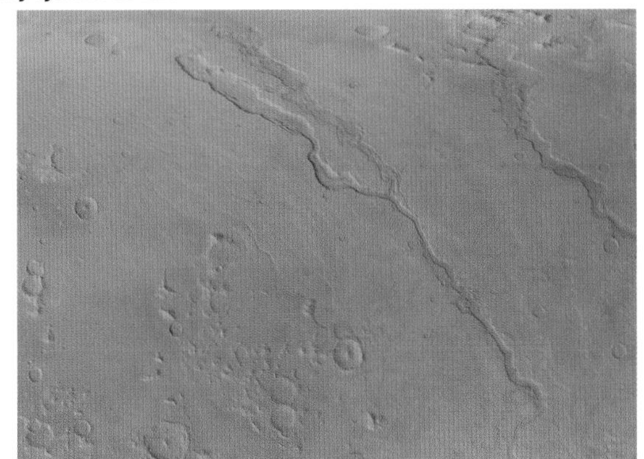

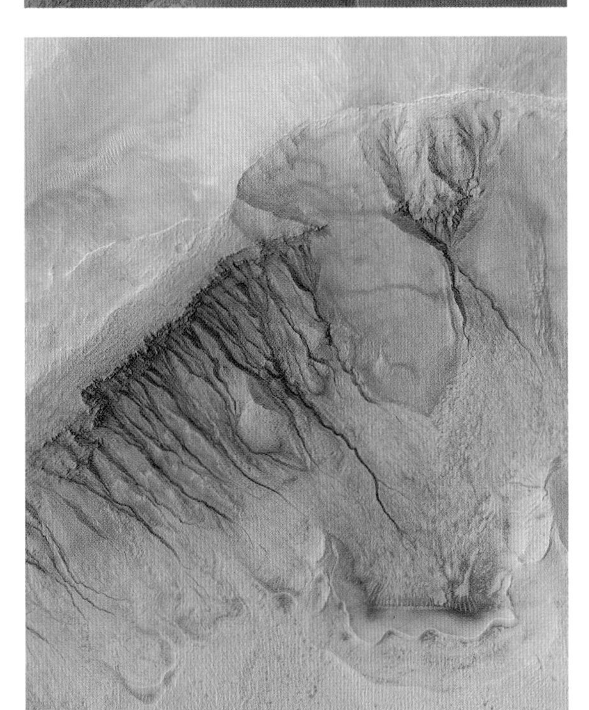

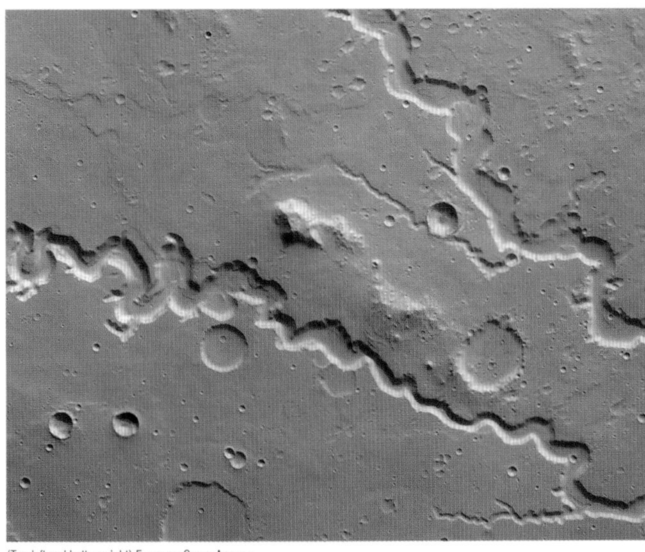

(Top left and bottom right) European Space Agency;
(top right and bottom left) NASA/JPL/Malin Space Science Systems

gullies that seemed to have formed relatively recently. These gullies appear on steep slopes in parts of the southern hemisphere. Some planetary scientists believe that the gullies were carved by water. They theorize that, episodically, small amounts of liquid water have flowed on and just below the surface in some places in geologically recent times. But other scientists disagree with this interpretation.

Interior. Scientists do not have direct information about the Martian interior. Instead, they develop models of the interior based on the planet's known characteristics, such as its size, mass, rotation rate, volcanic activity, magnetic properties, and gravity signature. In addition, more than 30 meteorites that have fallen to Earth are known to have come from Mars. The chemical composition of the meteorites indicates that Mars has separated into three main layers like Earth. Earth and Mars both have a metal-rich core at the center; a large, rocky middle layer called the mantle; and an outer crust.

Mars's core is probably rich in iron and sulfur. Scientists estimate that the core has a diameter of about 1,600–2,400 miles (2,600–4,000 kilometers). Unlike most other planets, Mars has no global magnetic field. Scientists believe that fluid motions in a planet's core help generate a magnetic field, so having such a field would indicate flow in the core. It is not known whether Mars's core is currently solid or liquid. However, ancient, highly magnetized rocks in its southern hemisphere suggest that in the past Mars had a strong magnetic field, which disappeared as the planet cooled. Scientists believe that Mars is still volcanically active, so its mantle is probably still warm and in some places is undergoing melting.

Measurements of the planet's gravity indicate that the crust is thinner and denser in the northern hemisphere than in the south. The thickness of the crust is thought to vary from about only 2 miles (3 kilometers) in places just north of the equator to more than 60 miles (90 kilometers) in southern Tharsis.

Moons

Mars's two moons, Phobos and Deimos, are small and rocky. They are named after sons of the ancient Greek war god Ares (the counterpart of the Roman god Mars). Phobos means "fear" in Greek, while Deimos means "terror." Both moons are so small that their gravity is too weak to pull them into spherical shapes. Instead, they are shaped more or less like potatoes. Phobos is about 16.5 miles (27 kilometers) long at its longest point, while Deimos is only about 9 miles (15 kilometers) long.

Each moon takes the same amount of time to rotate once on its axis as it does to complete one orbit around Mars. This means that, like Earth's moon, they always point the same face toward their planet. Deimos takes nearly one and a half Earth days to circle Mars, while Phobos completes about three orbits around Mars in one Earth day. Phobos is very close to Mars, and the planet's gravity draws it ever so slightly closer with each orbit. Astronomers think that Phobos might crash into Mars sometime in the next 100 million years. Deimos is in a more distant orbit and is gradually moving away from the planet.

(Top) European Space Agency; (bottom) National Aeronautics and Space Administration/Malin Space Science Systems

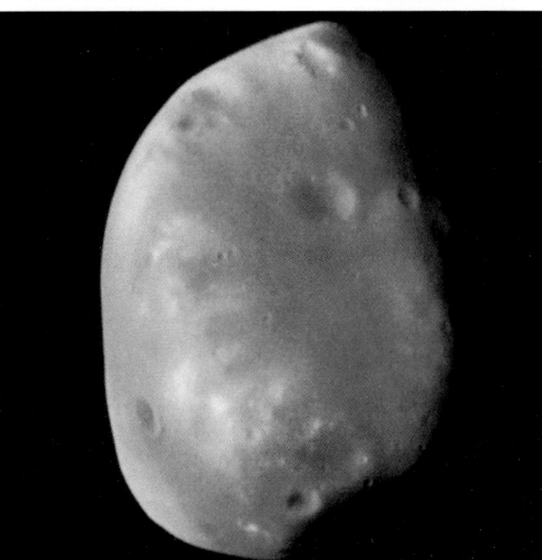

Mars has two small moons, Phobos (top) and Deimos (bottom). The surface of Phobos is grooved, pitted, and heavily cratered, while the surface of Deimos is smoother. Phobos appears in an image taken by Mars Express, with the large Stickney crater visible at left. Deimos, the smaller and outer moon, appears in an image taken by the Viking orbiters. The images are not to scale.

The surface of Phobos is very heavily cratered and grooved. One of its craters, named Stickney, is about half as wide as Phobos itself. The surface of Deimos appears smoother because its craters are almost buried in a layer of fine rubble. The moons reflect very little light. They are probably similar in composition to carbonaceous chondrite meteorites. Phobos and Deimos may once have been asteroidlike objects that came too close to Mars and were captured by its gravity.

Observation and Exploration

For centuries astronomers have considered the possibility that life might exist on Mars, the most Earth-like of the planets. In 1877 the Italian astronomer Giovanni Schiaparelli described what he believed was a system of interconnecting, straight-edged channels on the planet. The American astronomer Percival Lowell popularized the idea that these features were canals that had been built by an advanced but dying Martian civilization. Most astronomers could see no canals, however, and many doubted their reality. The controversy was finally resolved only when photographs taken by the Mariner space probes showed many craters but nothing resembling manufactured canals.

Many unmanned spacecraft have been sent to study Mars, including craft that have flown by, orbited, and landed on the planet. Wheeled robotic roving vehicles called rovers have also investigated the surface. No manned crews have yet been sent to the planet, but it is likely that Mars will be the first world other than Earth and Earth's moon that humans visit.

Four of the Mariner series of unmanned space probes launched by the United States National Aeronautics and Space Administration (NASA) investigated Mars. The first craft to successfully fly by Mars was Mariner 4, which photographed the planet as it passed by in 1965. Its images showed heavily cratered surfaces that resemble Earth's moon. Mariners 6 and 7 analyzed the atmosphere and captured images as they flew by Mars in 1969. The first spacecraft to orbit a planet other than Earth was Mariner 9. It photographed the Martian surface for nearly a year in 1971–72, revealing widespread volcanic activity and features carved by water in the remote past.

The Soviet Union also sent a series of unmanned space probes to Mars in the 1960s and '70s. Its Mars 3 lander was the first craft to successfully soft-land on the planet, in 1971. Unfortunately, it touched down during a global dust storm, which caused its communications systems to fail after about 20 seconds.

NASA's Viking probes consisted of two orbiting spacecraft and two landers. They were intended in part to search for evidence of past or present forms of life on Mars. The two landers touched down on the planet in 1976 and performed numerous experiments, including detailed chemical analyses of the Martian atmosphere and soil. No trace of complex organic material was found. A couple of experiments returned results that could have been caused by biological processes, but most scientists believe the results are better explained by nonbiological processes. In other words, though some of the results were inconclusive, they turned up no convincing signs of life on the surface near the landing sites.

NASA/JPL/Caltech

A detail of a composite of several images taken by the Mars Pathfinder lander shows the boulder-strewn surface at Chryse Planitia. In the distance are two hills dubbed "Twin Peaks," which are about 100 feet (30 meters) tall.

The Soviet Union sent two probes, Phobos 1 and 2, to study the Martian moon Phobos in 1988. Mission scientists lost contact with the first craft before it reached its target, but Phobos 2 successfully reached the moon in 1989. It collected data on Phobos and Mars for several days before it, too, malfunctioned.

A fairly high percentage of missions to Mars have failed, including three United States missions in the 1990s: Mars Observer, Mars Climate Orbiter, and Mars Polar Lander. Nozomi, a Japanese orbiter launched in 1998, reached Mars but then malfunctioned and could not be placed into orbit.

Moons of Mars

	Phobos	Deimos
average distance from center of Mars	5,827 miles (9,378 kilometers)	14,577 miles (23,459 kilometers)
dimensions	16.5 x 13.8 x 11.6 miles (26.6 x 22.2 x 18.6 kilometers)	9.3 x 7.6 x 6.5 miles (15 x 12.2 x 10.4 kilometers)
orbital period	0.3 Earth day	1.3 Earth days
rotation period	synchronous (same as orbital period)	synchronous (same as orbital period)
year of discovery	1877	1877
discoverer	Asaph Hall	Asaph Hall

Exploring Mars: Major Spacecraft Missions

Spacecraft	Type of Mission	Launch Date	Country or Agency	Highlights
Mariner 4	flyby	Nov. 28, 1964	U.S.	First spacecraft to fly past Mars, on July 14–15, 1965. Returned first spacecraft photographs of the surface.
Mariner 6	flyby	Feb. 24, 1969	U.S.	Flew by planet on July 31, 1969. Photographed Martian surface and measured composition of atmosphere.
Mariner 7	flyby	March 27, 1969	U.S.	Flew by planet on Aug. 5, 1969. Twin of Mariner 6.
Mars 3	orbiter and lander	May 28, 1971	U.S.S.R.	Lander was first craft to soft-land on Mars, on Dec. 2, 1971. Transmitted data for only 20 seconds owing to dust storm. Orbiter photographed planet and collected data on atmosphere.
Mariner 9	orbiter	May 30, 1971	U.S.	First spacecraft to orbit another planet. Entered orbit around Mars on Nov. 14, 1971, and operated until Oct. 27, 1972. Returned about 7,330 images, mapping about 80 percent of the surface, as well as analyzed planet's atmosphere, surface temperature, and gravity.
Viking 1	orbiter and lander	Aug. 20, 1975	U.S.	Lander touched down on Chryse Planitia on July 20, 1976. Transmitted first photographs from the surface of Mars and measured properties of atmosphere and soil. Tested soil to detect organic material and other evidence of life; some results inconclusive, but no convincing signs of life found. Orbiter mapped and analyzed large portions of surface.
Viking 2	orbiter and lander	Sept. 9, 1975	U.S.	Lander touched down on Utopia Planitia on Sept. 3, 1976. Twin of Viking 1.
Phobos 2	orbiter and attempted landers	July 12, 1988	U.S.S.R.	One of two spacecraft sent to study Phobos and Mars. Orbiter investigated targets for several days in 1989, but contact was lost before landers were to be deployed.
Mars Global Surveyor	orbiter	Nov. 7, 1996	U.S.	Began orbiting Mars on Sept. 12, 1997. Carried out long-term studies of atmosphere, surface, and aspects of interior. Produced detailed topographic maps of surface and high-resolution photos. Some showed gullies that may have been formed by water in geologically recent times. Completed primary mission in 2001. Extended mission continued until contact was lost on Nov. 2, 2006.
Mars Pathfinder	lander	Dec. 4, 1996	U.S.	Landed on Chryse Planitia on July 4, 1997. First craft to bounce to a landing on a cushioning cluster of air bags. Deployed rover named Sojourner, which took photographs and measured composition of rocks and soil. The data indicated that Mars was once warmer and wetter. Lander transmitted last data on Sept. 27, 1997.
Mars Odyssey	orbiter	April 7, 2001	U.S.	Reached planet on Oct. 24, 2001. Mapped chemical composition of surface. Confirmed water ice in subsurface. Completed primary mission in 2004 and began extended mission.
Mars Express	orbiter	June 2, 2003	ESA	Entered orbit on Dec. 15, 2003. Deployed British lander named Beagle 2, but lander failed to establish contact. Orbiter mapped atmosphere, surface, and subsurface and took high-resolution photos with stereoscopic camera.
Spirit (Mars Exploration Rover)	rover	June 10, 2003	U.S.	Landed in Gusev Crater on Jan. 4, 2004. Traveled over surface, taking photographs and studying properties of rocks and soil. Found mineral composition of rocks indicating presence of liquid water in the past. Completed original 90-day mission and began extended mission.
Opportunity (Mars Exploration Rover)	rover	July 8, 2003	U.S.	Landed in Meridiani Planum on Jan. 25, 2004. Twin of Spirit. Completed original 90-day mission and began extended mission.
Mars Reconnaissance Orbiter	orbiter	Aug. 12, 2005	U.S.	Arrived on March 10, 2006, to study planet's climate and surface features formed by water and to identify potential landing sites for future missions. Took very high-resolution images. Found vast deposits of clays, suggesting water in ancient times, and of what may be opal, suggesting water less than 2.5 billion years ago.
Phoenix	lander	Aug. 4, 2007	U.S.	Landed in north polar region on May 25, 2008. Collected soil with robotic arm and analyzed samples in onboard chemistry lab and furnace. Found subsurface water ice. Soil was alkaline and contained calcium carbonate, suggesting the past presence of water. Detected snowfall, with snow vaporizing before it could reach the ground.

Artists' conception of Mars Exploration Rover courtesy of NASA Jet Propulsion Laboratory

In addition to sending spacecraft to Mars, scientists also study meteorites that have fallen to Earth from the red planet. In 1996 a team of scientists announced that an ancient Martian meteorite contains organic matter and structures that resemble fossils of microscopic life-forms on Earth. The team believed that this provided the first evidence of life on early Mars. Most other scientists have been skeptical of that analysis. There is more widespread agreement that minerals in the meteorite were deposited there by liquid water.

In 1997 NASA's Pathfinder spacecraft landed on Mars to study the planet's geology and atmosphere. On board was a rover called Sojourner that took photographs and collected and analyzed samples of the Martian soil. Data gathered during Pathfinder's 83 days of surface operations indicated that Mars was once more Earth-like, with a thicker atmosphere and warmer temperatures. In addition, it found surface features that were probably formed by flowing water.

In 1996 NASA launched Mars Global Surveyor, the first in a series of orbiters designed to study the planet over longer periods. It began to orbit Mars in 1997. It mapped a variety of Mars's properties, including its gravity and magnetic fields and the surface topography and mineral composition, and took more than 200,000 images. Mars Global Surveyor collected more data about the planet than all other previous missions combined. Contact with the craft was lost in 2006.

Small metallic spheres, dubbed "blueberries," on Mars suggest the past presence of liquid water on the surface. Numerous spheres were found embedded in rock in Meridiani Planum near the landing site of the Mars Exploration Rover Opportunity. The rover analyzed a group of spheres (top) that had collected on a rock in "Eagle Crater." (The large, lighter circle on the rock surface was made by one of Opportunity's tools.) The rover found that the spheres are rich in hematite, a mineral that on Earth often forms in the presence of water. Scientists think that the spheres are concretions that formed out of minerals in water flowing through rocks; when the rocks later eroded, the spheres fell out. A microscopic view (bottom) taken by Opportunity shows similar round "blueberries" amid other pebbles in soil elsewhere in the crater. The spheres are actually about a fifth of an inch (4 millimeters) across.

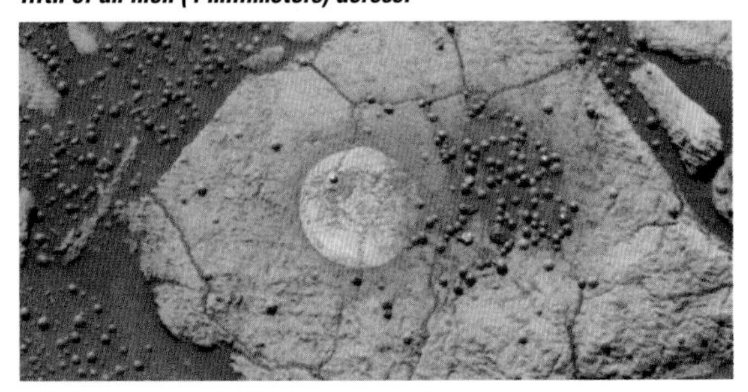

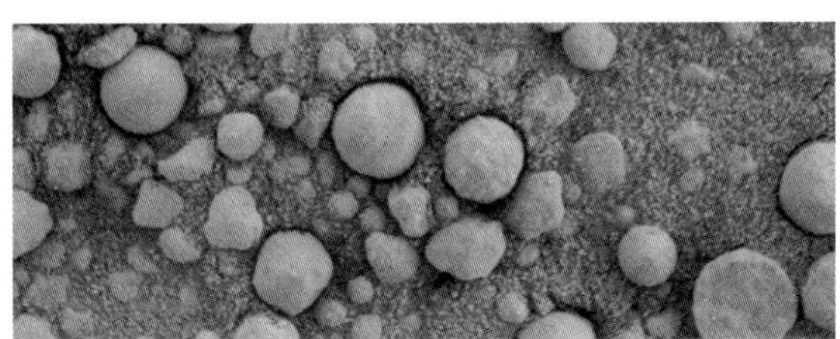

(Top) NASA/JPl/Cornell; (bottom) NASA/JPL/Cornell/USGS

NASA/JPL/Cornell

A detail of the "McMurdo" panorama taken by the Mars Exploration Rover Spirit shows the area around "Low Ridge," a hill in Gusev Crater on which the rover wintered in 2006. The rover's tracks are visible at left.

Several craft began exploring Mars in the early 21st century. NASA's global mapping orbiter Mars Odyssey reached the planet in 2001. In addition to mapping the chemical composition of the surface, it confirmed the presence of water ice just below the surface. Another mapping orbiter, NASA's Mars Reconnaissance Orbiter, arrived at the planet in 2006 to study the climate and surface features associated with liquid water.

The European Space Agency (ESA) sent its first mission to Mars in 2003. After arriving at the planet in December of that year, the orbiter Mars Express mapped a variety of properties in the Martian atmosphere, surface, and subsurface and photographed surface features using a high-resolution stereoscopic camera. It found what appears to be a large frozen sea just under the Martian surface near the equator. It also detected auroras, tiny amounts of methane in the atmosphere, and both water ice and carbon dioxide ice in the southern polar ice cap. The orbiter carried on board a British lander named Beagle 2. After being released at Mars, however, the lander failed to return communications signals and was declared lost.

NASA's twin Mars Exploration Rovers, named Spirit and Opportunity, landed on Mars in 2004. The rovers collected geologic data to help determine whether the planet's environment was suitable for life in the past. Spirit landed in Gusev Crater, while Opportunity arrived at Meridiani Planum. Continuing to function years after their expected demise, they traversed several miles each, even both surviving a severe dust storm in 2007. Opportunity explored the large crater called Victoria from 2006 to 2008 before heading to Endeavour crater. Each rover was equipped with several cameras and instruments, including a microscopic imager and a rock-grinding tool. Instruments called spectrometers allowed them to detect the patterns of heat and other radiation emitted by the rocks. With these data, scientists determined the rocks' mineral compositions. The rovers found several clues suggesting the past existence of liquid water on the surface, including what seems to have been an ancient body of salty water.

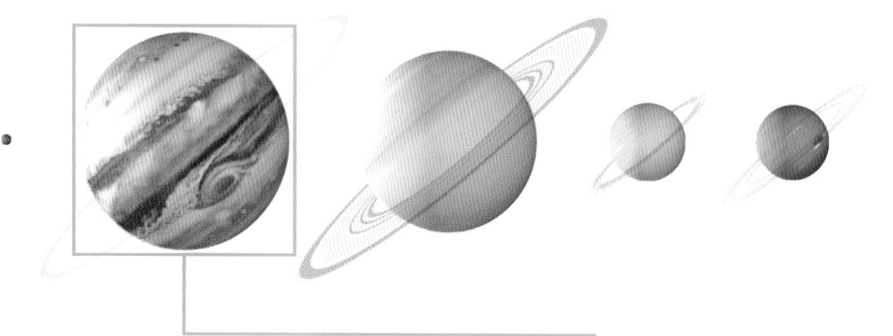

JUPITER

The fifth planet from the sun and the solar system's largest planet by far is Jupiter. More than 1,300 Earths would fit inside it. The planet is one of the brightest objects in the night sky, and even a small telescope can reveal its multicolored stripes. These stripes are bands of clouds being pushed around the planet by strong east-west winds. Jupiter is a world of complex weather patterns. Its most prominent feature is an orange-red oval called the Great Red Spot. The oval is a storm system that has lasted at least 300 years and is bigger across than Earth and Mars combined.

Jupiter is not just bigger than Earth, it is also fundamentally different in composition. Jupiter has no solid surface. It is formed of the same elements, in roughly the same proportions, as the sun and other stars. Like Saturn, it is made almost entirely of hydrogen and helium in liquid and gaseous forms. Although Jupiter is huge, it would have to be about 80 times more massive in order to generate nuclear reactions and become a star.

Nevertheless, Jupiter reigns at the center of a system of dozens of moons like a miniature solar system. Four of the moons are quite large and would probably be considered planets themselves if they did not orbit a planet. They are remarkably dissimilar worlds. Io is the most volcanically active body in the solar system. Frequent eruptions renew its surface and cover over any craters that form. Europa's surface is also fairly young, but its features are smoothed not by fiery lava but by water ice, which possibly flows up from beneath the surface and freezes. Europa may harbor an ocean of warm, salty water just under its icy crust. Callisto has an ancient crater-scarred surface that seems to have been largely undisturbed by geologic activity for some 4 billion years. Ganymede, the solar system's largest moon, is bigger than the planet Mercury and has a planetlike magnetic field.

NASA/JPL/University of Arizona

Bands of pastel-colored clouds encircle the giant planet Jupiter, which appears in a composite of images taken by the Cassini spacecraft. The small black disk at lower left is a shadow cast by Jupiter's moon Europa. The enormous storm called the Great Red Spot is visible at lower right. The raw images were taken in two colors, which were processed to simulate Jupiter's natural colors.

Basic Planetary Data

An outer planet, Jupiter is much farther from the sun than Earth and the other inner planets are. Its orbit lies between the main asteroid belt and Saturn's orbit. Its inner planetary neighbor, Mars, orbits between the asteroid belt and Earth. Like the other outer planets—Saturn, Uranus, and Neptune—Jupiter is much larger and less dense than Earth and the other rocky inner planets.

Size, mass, and density. Jupiter is named for the ruler of the ancient Roman gods, the equivalent of the ancient Greek god Zeus. The ancient Romans did not know how large the planet is, but the name turned out to be fitting. Jupiter encompasses more matter than all the other planets in the solar system combined. Its diameter at the equator is some 88,846 miles (142,984 kilometers). It is about 320 times as massive as Earth. Jupiter's large mass produces strong gravitational effects on other members of the solar system. It forms gaps, for instance, in the distribution of the asteroids in the main belt and changes the trajectory of comets. The planet's density is very low, only about 1.3 times that of water. By comparison, Earth is more than 5.5 times denser than water.

NASA/JPL

Ribbons of clouds wind through the area around Jupiter's huge Great Red Spot, an ancient, whirling storm system, in an image captured by Voyager 1. Below the Great Red Spot is another large whirling storm, which appears as a white oval.

Facts About Jupiter

Average Distance from Sun. 483,000,000 miles (778,000,000 kilometers).

Diameter at Equator.* 88,846 miles (142,984 kilometers).

Average Orbital Velocity. 8.1 miles/second (13.1 kilometers/second).

Year on Jupiter.† 11.86 Earth years.

Rotation Period.‡ 9.9 Earth hours.

Day on Jupiter (Solar Day). 9.9 Earth hours.

Tilt (Inclination of Equator Relative to Orbital Plane). 3.1°.

Atmospheric Composition. Mostly hydrogen, with some helium, a small amount of methane, and trace amounts of ammonia, water vapor, other gases.

General Composition. Hydrogen, helium.

Weight of Person Who Weighs 100 pounds (45 kilograms) on Earth.* 235.9 pounds (107 kilograms).

Number of Known Moons. 63.

*Calculated at the altitude at which 1 bar of atmospheric pressure (the pressure of Earth's atmosphere at sea level) is exerted.

†Sidereal revolution period, or the time it takes the planet to revolve around the sun once, relative to the fixed (distant) stars.

‡Sidereal rotation period, or the time it takes the planet to rotate about its axis once, relative to the fixed (distant) stars.

Orbit and spin. Jupiter, like all the other planets, travels around the sun in a slightly elliptical, or oval-shaped, orbit. It completes one orbit in about 11.86 Earth years, which is the length of a year on Jupiter. Its average distance from the sun is about 483 million miles (778 million kilometers), which is more than five times greater than Earth's.

Jupiter spins very quickly on its axis, faster than the other seven planets. It completes one rotation in about 9 hours and 55 minutes, which is the length of a day on Jupiter. The atmosphere spins at slightly different rates, with the clouds near the equator completing a rotation a few minutes faster than the clouds at higher latitudes. The force of a planet's rotation causes it to bulge slightly at the equator and to flatten slightly at the poles. Jupiter's rapid spin accentuates this, so it is less perfectly spherical than most of the other planets. Jupiter's spin axis is hardly tilted at all. For this reason, it does not have seasons like Earth, Mars, and other planets with tilted axes.

Atmosphere

Jupiter has a very massive atmosphere, or layer of surrounding gases. It is about 86 percent hydrogen and 14 percent helium by mass. The sun has a similar composition, at about 71 percent hydrogen and 28 percent helium by mass. Planetary scientists believe that the four outer planets all received about the same proportions of hydrogen and helium as the sun when the solar system formed from a disk of gas and dust (*see* Planet). It is thought that in Jupiter and Saturn more of the helium is concentrated in the interior. Jupiter's atmosphere also contains trace amounts of many other gases, including methane, ammonia, water vapor, hydrogen sulfide, and hydrogen deuteride.

In the lower parts of the atmosphere, where clouds form, it generally gets colder with increasing height above the planet. At higher levels, the gases and particles absorb solar radiation. This makes the middle and upper levels of the atmosphere hotter, with temperatures increasing with altitude.

The clouds in Jupiter's atmosphere appear as alternating dark and bright bands roughly parallel to the equator. The darker bands are called belts, while the brighter bands are called zones. The clouds are also separated into different layers by depth. They range in color from white to tawny yellow, brown, salmon, and blue-gray. Scientists think that the clouds vary in color because they contain different chemicals.

The highest clouds are white and are composed of frozen crystals of ammonia. The temperature at their tops is about –240° F (–150° C). Clouds in the main deck are tawny colored and lower in the atmosphere, where the temperature is about –100° F (–70° C). In some places holes in the layers of tawny clouds reveal

A map shows the entire visible "surface" of Jupiter, including its bands of clouds, the Great Red Spot, and many smaller oval storms. The map is a cylindrical projection, like maps of Earth that show the planet stretched out flat onto a rectangle. It was generated by computer from 36 images taken by the Cassini spacecraft in two colors; the images were processed to simulate the planet's true colors. Jupiter's polar regions appear less clear because of the angle of the spacecraft's camera and because they were covered by a thicker haze.

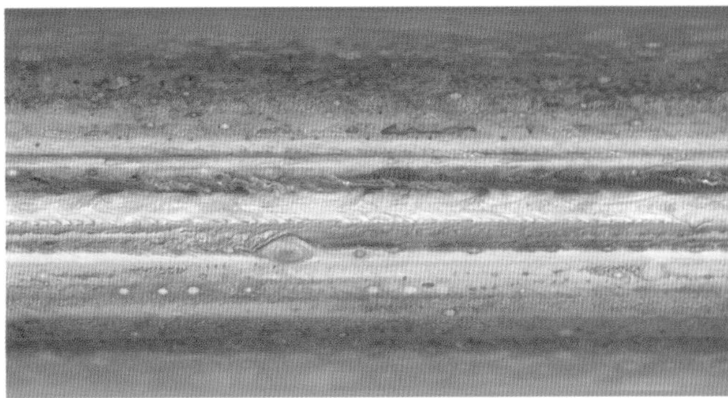

NASA/JPL/Space Science Institute

dark brown clouds below. The tawny and brown clouds are probably made mostly of ammonium hydrosulfide, and their colors may result from other sulfur compounds. Scientists also think there is a lower deck of clouds formed of water ice and water droplets. Blue-gray and purplish areas are found only near the equator. They are thought to be areas with relatively few or no clouds.

Jupiter has a turbulent atmosphere, and its cloud systems form and change in a matter of hours or days. However, the underlying pattern of wind currents has been stable over decades. Strong winds blow east or west through the atmosphere in several alternating bands. They are interrupted in places by large whirling storm systems that appear from above as ovals.

The most persistent feature in the atmosphere is the famous Great Red Spot. It has been observed from Earth since 1664. The spot is a huge oval-shaped storm system in the planet's southern hemisphere, with strong winds swirling counterclockwise about a high-pressure center. In other words, it is like an anticyclone on Earth. It covers an area larger than Earth itself, however, with dimensions of about 12,400 by 7,500 miles (20,000 by 12,000 kilometers). Material within the spot completes one circle about every seven days. This means that winds around the outer parts of the storm probably reach super-hurricane force, blowing at some 250 miles (400 kilometers) per hour. As Jupiter rotates, the storm system moves in longitude with respect to the clouds, but it remains centered at about latitude 22° S.

The Great Red Spot projects higher than the planet's highest white clouds, and it probably also descends well below the main cloud layers. There is no clear evidence that the storm causes upwelling of material at its center, though some vertical movement would be expected. Scientists are not certain why the spot is reddish. They think its color might result from complex organic molecules, red phosphorous, or sulfur compounds. Any of these materials could be produced by lightning. They also could result from material upwelling to high altitudes, where it reacts chemically with ultraviolet radiation from the sun.

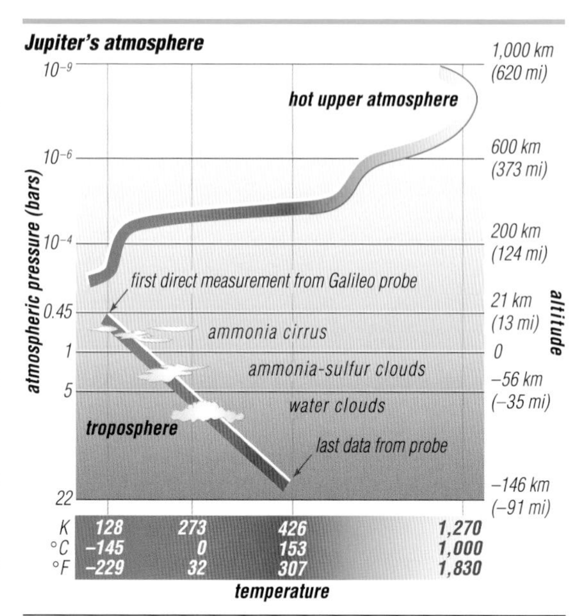

A graph shows the temperatures and pressures at different levels of Jupiter's atmosphere, as determined by the Galileo probe. Pictures of clouds indicate the approximate positions of the expected cloud layers; the probe did not detect them, having descended in a nearly cloudless spot. Some of the temperatures and pressures were directly measured by the probe, while others were deduced from other data collected by it.

As of 2006, Jupiter had two large red spots. The Great Red Spot, below the equator at right, and the much younger red storm nicknamed Red Spot, Jr., at lower center, appear in an image taken by the Hubble Space Telescope in April 2006. The smaller red spot is about as big across as Earth. The false-color image was taken in visible light and at near-infrared wavelengths.

NASA/ESA

Three smaller, white, oval-shaped storms were observed just south of the Great Red Spot starting in about the 1940s. The three ovals merged in 1998–2000, creating a single storm system nearly half as big across as the Great Red Spot. Scientists believe that a similar merger may have created the Great Red Spot. In fact, in 2006 the merged storm turned the same salmon-red as the larger spot, for reasons that are not yet known.

Interior

The atmosphere surrounding Jupiter makes up only a small percentage of the planet. Scientists cannot directly observe the planet below the atmosphere, however. Instead, they form theoretical models based on many known properties such as the planet's size, mass, density, rotation rate, heat balance, and atmospheric pressures and temperatures.

Like the atmosphere, the interior is composed mainly of hydrogen and helium. Inside the planet, pressures and temperatures increase greatly with depth, so the hydrogen and helium get denser and denser. Starting at about a quarter of the way down to the center, the pressure has probably squeezed the hydrogen into liquid metallic form. In this state, the electrons are stripped away from the atomic nuclei, so the fluid hydrogen would conduct electricity like a metal.

At Jupiter's center is probably a very dense core. Different models have the core about a third as big as Earth to a bit bigger than Earth. Temperatures there may reach nearly 45,000° F (25,000° C). The pressure in the core is likely 50 million to 100 million times the pressure at sea level on Earth.

Jupiter has some sort of internal heat source. The planet emits nearly twice as much energy as it receives from the sun, for reasons that are not entirely clear. Much of this heat was probably acquired during the planet's formation some 4.6 billion years ago. As the planet continues to cool off, it gradually emits heat. Scientists think that another process probably also generates some of the heat. This process involves helium separating out into droplets and sinking toward the planet's center. The friction of the helium droplets pushing against the liquid metallic hydrogen would convert some energy to heat.

Magnetic Field and Magnetosphere

Jupiter has the largest and strongest magnetic field of all the planets. The planet's rapidly rotating, electrically conducting interior is thought to give rise to the strong field. Like Earth's magnetic field, it has two poles, north and south, like a giant bar magnet. The orientation of the poles is opposite that of Earth, so that a compass would point south on Jupiter. Jupiter's magnetic field is also tilted about 10 degrees relative to its spin axis.

The region of space dominated by Jupiter's magnetic field is called its magnetosphere. It is a huge teardrop-shaped area. On the side nearest the sun it extends about 1.9 million miles (3 million kilometers). On that side, the magnetosphere holds off the solar wind, which is a flow of electrically charged particles from the sun. This creates a large shock wave. On the opposite side of Jupiter, the solar wind pushes the magnetosphere's tail out to the orbit of Saturn, some 400 million miles (650 million kilometers) away.

Jupiter's magnetic field traps electrically charged particles around the planet. The particles move around Jupiter in roughly doughnut-shaped regions. Electrons traveling almost at the speed of light radiate energy as they spiral through the regions. These regions of intense radiation are similar to but stronger than Earth's Van Allen radiation belts (*see* Earth, "Magnetic Field and Magnetosphere").

Photo AURA/STScI/NASA/JPL (NASA photo # PIA01257, STScI-PRC96-32)

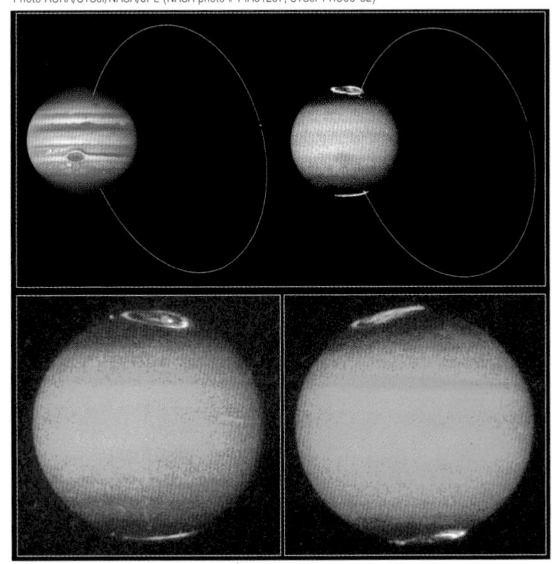

Auroras light up Jupiter's poles in images taken by the Hubble Space Telescope. The two lower images are in false color and were taken in ultraviolet light. They follow changes in the brightness and structure of the auroras as the planet rotates. In the two top images, a line was added to trace the path of the magnetic flux tube, or current of charged particles, that links Jupiter and its moon Io. The image at top left was taken in visible light, while the one at top right was taken in ultraviolet light.

Jupiter strongly emits radio waves in both intermittent bursts at longer wavelengths and steady streams at shorter wavelengths. Both types result from charged particles moving in the planet's magnetosphere. The bursts are sometimes the most intense source of radio "noise" in the sky. They were first detected in the 1950s, and they provided the first clues that Jupiter had a magnetic field. The steady emissions, which were discovered later, are radiated by the charged particles trapped in the radiation belts. This stream of radio waves varies somewhat in intensity and orientation as the planet rotates. The variations have a characteristic period, which is the rotation rate of Jupiter's magnetic field. The rotation rate of the magnetic field is also the rotation rate of the planet's interior, which produces the field.

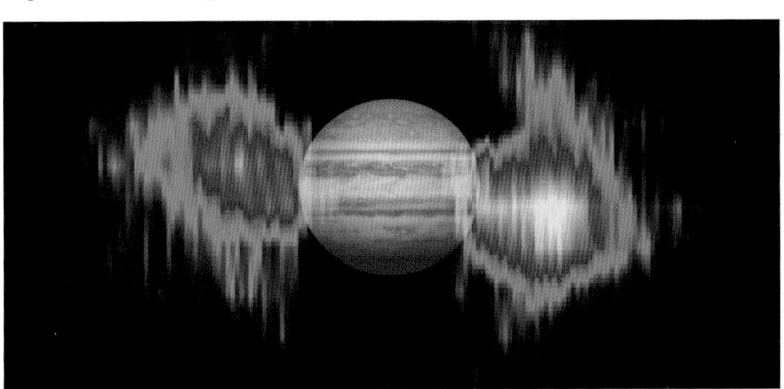

The Cassini spacecraft mapped Jupiter's radiation belts by measuring the strength of their radio emissions at a frequency of 13,800 megahertz (13.8 billion cycles per second). Color coding indicates the strength of the emission, with yellows and reds being the most intense. A photograph of Jupiter taken by telescope has been added to show the size and orientation of the belts relative to the planet.

NASA/JPL

The bursts of radio waves come from three distinct areas around Jupiter. The position of the moon Io as it orbits Jupiter is thought to strongly influence these bursts. Magnetic field lines connect Jupiter to Io. They enclose a doughnut-shaped region of space called a flux tube between the planet and the moon. This flux tube moves along with Io. In addition, volcanic eruptions on Io release a cloud of electrically charged particles that accompanies Io along its orbit. As the cloud passes through Jupiter's magnetic field, an electric current of some 5 million amperes is generated. Scientists believe that the radio bursts are probably emitted by electrons that spiral along the magnetic field lines in the flux tube connecting Io and Jupiter.

Auroras similar to Earth's northern and southern lights appear at times near Jupiter's poles. As on Earth, the auroras result from charged particles in the radiation belts crashing into molecules of the upper atmosphere.

Ring System

Jupiter's thin ring system was discovered only by spacecraft, by Voyager 1 in 1979. The rings are composed of tiny dust particles that orbit the planet. The main ring is about 4,000 miles (6,400 kilometers) wide and 19 miles (30 kilometers) thick. Its outer edge lies some 80,000 miles (129,000 kilometers) from the planet's center. Straddling the main ring are an inner cloudlike ring of particles called the halo and two outer rings. The outer rings are called gossamer rings because the particles within them are very thinly distributed. All the rings are formed of debris produced when small fragments of asteroids, comets,

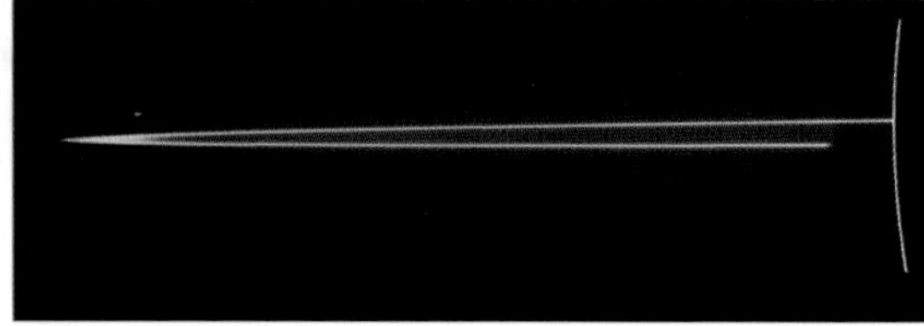

NASA/JPL

An image taken by the Galileo orbiter shows the thin main ring of Jupiter nearly edge-on, in natural light. The nearer arm of the ring disappears close to Jupiter where it passes into the planet's shadow.

and other objects collide with Jupiter's four small inner moons. The main ring seems to be formed of debris from the moons Metis and Adrastea. The gossamer rings are fed by Amalthea and Thebe.

Moons

More than 60 known moons orbit Jupiter, and more are likely still to be discovered. The four biggest moons are each about the size of Earth's moon or larger. In order of increasing distance from Jupiter, they are Io, Europa, Ganymede, and Callisto. They were the first objects in the solar system to be discovered with a telescope, by Galileo in 1610. They are now called the Galilean satellites in his honor. Amalthea was the next of Jupiter's moons to be discovered. It was first observed

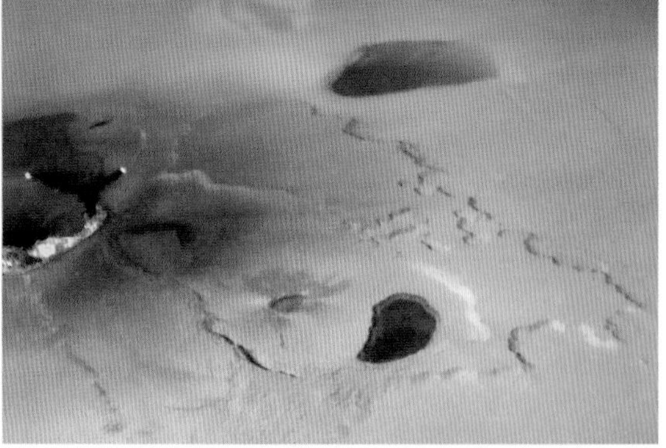

NASA/JPL/California Institute of Technology

Hot, glowing lava erupts from a volcano on Io in an image captured by the Galileo orbiter. It is a mosaic of several images, including enhanced-color images taken in visible light. The bright white, yellow, and orange at left are false colors added to images taken at infrared wavelengths to indicate temperature, with white being the hottest and orange the coolest.

by Edward Emerson Barnard in 1892 through a telescope. All the other moons of Jupiter were discovered by examining images captured by Earth-based telescopes or the Voyager spacecraft. Many of the planet's smaller moons are less than 5 miles (8 kilometers) in diameter and were discovered in 2000 or later, with powerful telescopes and highly sensitive electronic imaging equipment.

Jupiter's eight inner moons—the four small moons Metis, Adrastea, Amalthea, and Thebe plus the four Galilean satellites—orbit the planet in fairly circular paths. The plane of their orbit is also nearly the same as the plane in which Jupiter orbits the sun. For these two reasons, the moons are called regular. The rest of Jupiter's moons are irregular. That is, they have a very elongated or tilted orbit, or both. Also, the regular moons orbit Jupiter in the same direction as the planet's rotation, while most of the irregular moons orbit in the opposite direction. Scientists believe that Jupiter's eight regular moons probably formed along with Jupiter some 4.6 billion years ago. As Jupiter formed from a disk of gas and dust surrounding the sun, its regular moons formed from such a disk surrounding the planet (*see* Planet). The irregular moons may have been asteroids, comets, or fragments that passed close enough to Jupiter for its gravity to capture them into orbits.

The two inner Galilean satellites, Io and Europa, are much denser than the outer two, Ganymede and Callisto. Io and Europa are probably rocky bodies with compositions roughly similar to Earth's moon. Ganymede and Callisto are probably roughly half rock and half water ice or some other substance of low density. The surfaces of three of the moons, Europa, Ganymede, and Callisto, are icy.

Io. Scientists think that Io has a molten core of iron and iron sulfide surrounded by a molten rocky middle layer, called a mantle. The moon's brightly colored surface is covered with volcanoes, lava, and deposits of sulfur in various forms. The sulfur gives Io its bright yellowish color, along with its smaller areas of red, orange, and black. The white areas on its surface are thought to be mostly solid sulfur dioxide.

Io is continually squeezed in and out by the strong gravitational pull of Jupiter and the weaker pull of the

NASA/JPL/Caltech

Jupiter's Galilean moons—from left to right, Io, Europa, Ganymede, and Callisto—appear in a montage created with images from the Galileo orbiter. The images are scaled to show the moons' sizes relative to one another.

other Galilean satellites because of the pattern of their orbits. This effect, called tidal flexing, generates internal friction and heat in Io. As a result, it is volcanically hyperactive. Data collected by the Galileo spacecraft suggest that about 300 volcanoes may be active on Io at any one time. The continual flow of lava completely replaces the surface every few thousand years. As discussed above, the volcanic eruptions release clouds of electrically charged particles that interact with Jupiter's magnetic field in complex ways.

Europa. Europa is especially intriguing because it is one of very few places in the solar system other than Earth that might have the liquid water necessary to support life. Like Io, it probably has an iron-rich core surrounded by a rocky mantle. Europa's bright icy surface, however, is the smoothest surface found on any known solid body in the solar system. There is very little variation in surface elevation. Interlacing grooves and ridges etch unusual and intricate patterns on the crust. There are hardly any large craters on Europa, which indicates that its surface has formed recently. (In general, the more craters a solid body has, the older it is.) This resurfacing may still be occurring. It is possibly the result of water flowing out from the interior and freezing on the crust.

Scientists believe that Europa may have a global ocean of liquid water roughly 60 miles (100 kilometers) deep, with the top several miles frozen. The water is believed to be warmed in the interior by tidal heating,

NASA/JPL/California Institute of Technology

A small, elaborately patterned area of Europa's ice crust appears in an enhanced-color image that combines images and data gathered by the Galileo spacecraft. Observations of such intricate structures on Europa indicate that in the recent past its crust cracked and huge blocks of ice rotated slightly before being refrozen in new positions. The size and geometry of the blocks suggest that their motion was enabled by an underlying layer of icy slush or liquid water present at the time of the disruption.

caused by the gravitational pull of Jupiter and the other Galilean moons. Images taken by the Galileo spacecraft seem to indicate that fluid motions have occurred at Europa's surface during the recent past. Blocks of ice seem to have fractured and moved relative to one another before refreezing. In addition, the moon disturbs Jupiter's magnetic field in a pattern that seems to be caused by an electrically conductive substance such as salt water.

Ganymede. Ganymede and Callisto are both cold worlds about the size of Mercury. Ganymede is actually a bit larger than that planet. It likely has an iron-rich core, a rocky lower mantle, and an upper mantle and

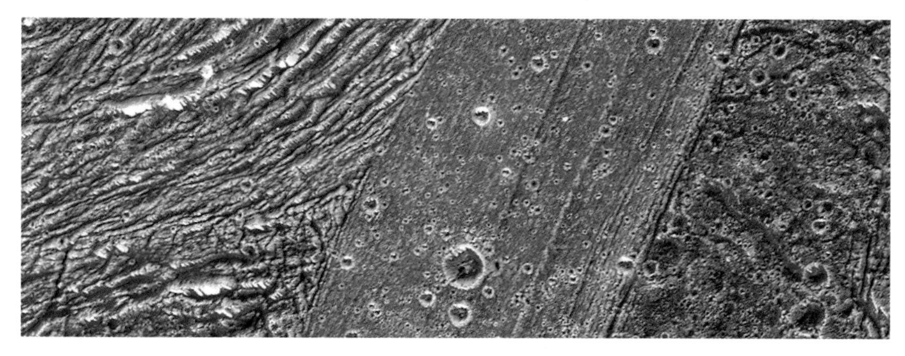

A close-up of Ganymede's surface taken by the Galileo spacecraft shows a region of diverse terrain about 55 miles (90 kilometers) long. The lightly cratered band cutting through the center is the youngest terrain. It divides the oldest terrain in the area, at right, from a grooved, highly deformed terrain that is intermediate in age, at left.

NASA/JPL/California Institute of Technology

crust of water ice. The moon probably has a deep ocean of liquid water underneath its icy crust.

The surface has two strikingly different kinds of terrain, dark and bright. The dark areas are heavily cratered and so are probably very old. The bright areas are younger and contain craters that have bright, radiating streaks of exposed water ice. Sets of long narrow grooves also mark the brighter terrain. This terrain formed at a point when Ganymede was geologically active, probably early in its history. The internal activity probably caused the crust to fracture into grooves.

Ganymede is the only moon in the solar system that produces its own permanent, strong magnetic field. Its magnetic field is about the strength of Mercury's and creates its own magnetosphere and auroras.

Callisto. Unlike the three other Galilean moons, Callisto seems to have only partially separated into layers. This suggests it never experienced as much tidal heating as did Io, Europa, and Ganymede. Tidal heating partly melted the interiors of those moons, and distinct layers formed. Instead, Callisto's interior seems to consist mostly of well-mixed rock and water ice.

NASA/JPL/Space Science Institute

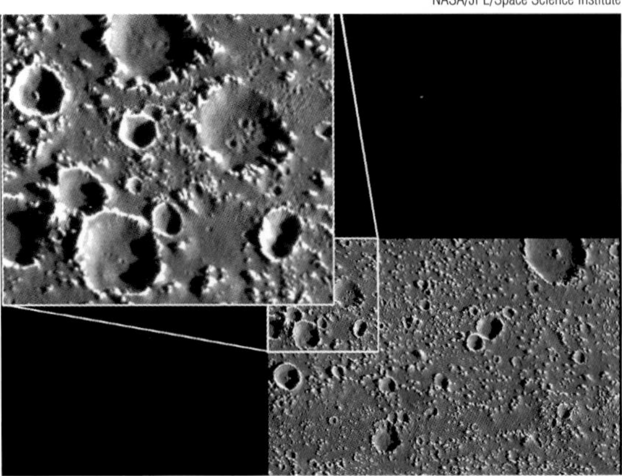

A heavily cratered and pitted region of Callisto's ancient surface appears in an image (bottom right) taken by the Galileo spacecraft and an enlargement (upper left) to show detail. The region shown in the bottom image is about 45 miles (72 kilometers) across.

Callisto's surface is quite dark and very heavily cratered. Little geologic activity seems to have occurred on its surface or interior for the past few billion years. Like Europa, Callisto disturbs Jupiter's

Exploring Jupiter: Major Spacecraft Missions

Spacecraft	Type of Mission	Launch Date	Country or Agency	Highlights
Pioneer 10	flyby	March 3, 1972	U.S.	First mission to Jupiter and outer part of solar system. Flew by the planet on Dec. 3, 1973. Along with Pioneer 11, measured radiation, magnetic field, properties of atmosphere and photographed and planet and moons. Discovered huge tail of magnetosphere.
Pioneer 11	flyby	April 6, 1973	U.S.	Flew past Jupiter on Dec. 4, 1974, on its way to study Saturn.
Voyager 2	flyby	Aug. 20, 1977	U.S.	First craft to visit all four outer planets. Along with Voyager 1, took advantage of rare orbital positioning of Jupiter, Saturn, Uranus, and Neptune that permitted multiplanet tour with relatively low fuel requirements and flight time. The twin craft's photographs and measurements revealed many previously unknown details of outer planets and their moons. Flew past Jupiter on July 9, 1979. Afterward visited other outer planets before being sent on path to take it out of solar system.
Voyager 1	flyby	Sept. 5, 1977	U.S.	Twin of Voyager 2. Flew by Jupiter on March 5, 1979, before heading for Saturn and then being sent on path to take it out of solar system.
Galileo	orbiter and atmospheric probe	Oct. 18, 1989	U.S.	First craft to orbit Jupiter and first probe to make contact with atmosphere of outer planet. First craft to fly near asteroids, Gaspra (1991) and Ida (1993). Discovered Dactyl, first moon found orbiting an asteroid, Ida. Orbiter released probe toward Jupiter on July 13, 1995. Probe parachuted below cloud tops for almost 58 minutes on Dec. 7, 1995, measuring many properties of atmosphere. Orbiter studied Jupiter and its major moons for nearly five years. After mission ended in 2000, sent to be destroyed in planet's atmosphere in 2003.
Cassini	flyby	Oct. 15, 1997	U.S.	Flew by Jupiter on Dec. 30, 2000, on way to primary mission at Saturn.

Artist's conception of Galileo orbiter courtesy of NASA/JPL/California Institute of Technology

magnetic field in a way that suggests that the moon contains a conductive substance such as salt water. It is possible that this moon, too, has some liquid salt water under its crust.

Spacecraft Exploration

The first three spacecraft missions to Jupiter—named Pioneer, Voyager, and Galileo—dramatically increased scientists' knowledge about the giant planet in the late 20th century. Two Pioneer spacecraft flew by Jupiter in the early 1970s to survey the planet's basic environment and assess whether its radiation levels would permit future spacecraft exploration. They were launched by the United States National Aeronautics and Space Administration (NASA). Pioneer 10 was the first spacecraft to travel beyond the asteroid belt to the outer part of the solar system. Flying within about 80,000 miles (130,000 kilometers) of Jupiter's cloud tops in 1973, it transmitted the first close-up images of the planet. It also discovered the huge tail of the planet's magnetosphere. Pioneer 11 followed, passing within about 27,000 miles (43,000 kilometers) of the cloud tops in 1974.

NASA's Voyagers 1 and 2 flew past Jupiter in 1979. Their instrumentation was more robust and sophisticated than the Pioneers', and they gathered much valuable data. Close-up images from the spacecraft also uncovered a few new moons, volcanic activity on Io, and a thin ring around Jupiter.

In 1989 NASA launched the Galileo spacecraft toward Jupiter for an extended study of the planet, its moons, and its magnetic field. When it reached the planet in 1995, it released a probe, which became the first man-made object to make contact with an outer planet. The probe parachuted through about 100 miles (165 kilometers) of the atmosphere, relaying measurements of the chemical composition, temperature, and pressure before being destroyed within about an hour by the planet's extreme conditions. Scientists expected the Galileo probe to detect water clouds below the main cloud deck, but it did not. Unfortunately, the probe seems to have dropped through a nearly cloudless area of the atmosphere.

The Galileo orbiter transmitted spectacular images and a wealth of data over several years, until it was nearly out of propellant. In 2003 it was intentionally sent on a collision course with Jupiter's atmosphere to destroy it.

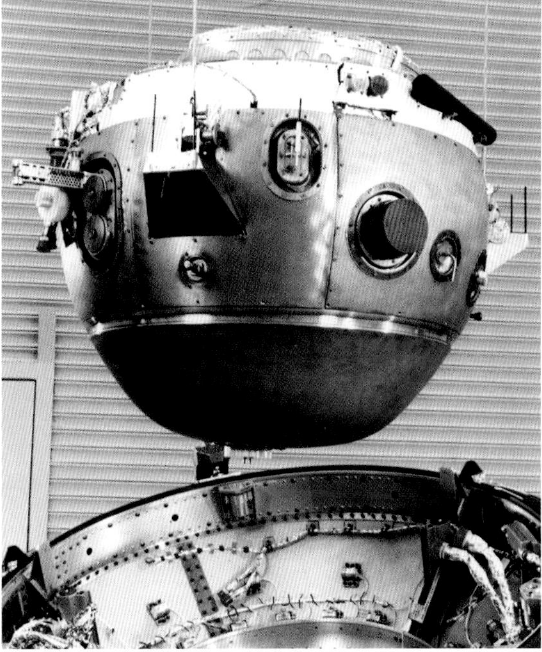

NASA/Ames Research Center

The Galileo atmospheric probe (shown before its launch) became the first man-made object to make contact with one of the outer planets. In 1995 the Galileo orbiter released the 747-pound (339-kilogram) probe on a course toward Jupiter. With the aid of a parachute, the probe slowly descended through the giant planet's atmosphere, using its six science instruments to measure such properties as temperature, pressure, density, cloud structure, and chemical composition.

A different opportunity to study Jupiter arose in 1994, when pieces of Comet Shoemaker-Levy 9's nucleus crashed into the planet's atmosphere. Scientists observed the effects of the explosions from Earth-based and Earth-orbiting telescopes and through images captured by the Galileo spacecraft. The impacts and the temporary black scars they formed in the planet's clouds provided clues about the composition and structure of Jupiter's atmosphere as well as about the comet's consistency.

Additional data and images of Jupiter were captured by NASA's Cassini spacecraft as it flew by the planet in 2000–01 on its way to Saturn. Among the phenomena studied through Cassini were large amounts of charged particles escaping from one side of Jupiter's magnetosphere.

Major Moons of Jupiter

	Io	Europa	Ganymede	Callisto
average distance from center of planet	262,000 miles (422,000 kilometers)	417,000 miles (671,000 kilometers)	665,000 miles (1,070,000 kilometers)	1,170,000 miles (1,883,000 kilometers)
diameter	2,260 miles (3,630 kilometers)	1,940 miles (3,130 kilometers)	3,275 miles (5,270 kilometers)	3,000 miles (4,800 kilometers)
density	3.53 grams per cubic centimeter	3.01 grams per cubic centimeter	1.94 grams per cubic centimeter	1.83 grams per cubic centimeter
orbital period	1.8 Earth days	3.6 Earth days	7.2 Earth days	16.7 Earth days
rotation period	synchronous (same as orbital period)	synchronous (same as orbital period)	synchronous (same as orbital period)	synchronous (same as orbital period)

NASA/JPL/Space Science Institute
The planet Saturn and its spectacular rings appear in a natural-color composite of 126 images taken by the Cassini spacecraft. The view is directed toward Saturn's southern hemisphere, which is tipped toward the sun. Shadows cast by the rings are visible against the bluish northern hemisphere, while the planet's shadow appears on the rings to the left.

SATURN

The sixth planet from the sun is Saturn. Dusty chunks of ice—some the size of a house, others of a grain of sand—make up its extraordinary rings. The other outer planets also have rings, but Saturn's are much larger and more complex. The planet is a popular target for amateur astronomers, because even a small telescope can reveal the dazzling rings. To the unaided eye, Saturn looks like a bright nontwinkling point of light. It was the most-distant planet known to ancient astronomers.

Saturn was named after the ancient Roman god of agriculture. His counterpart in ancient Greek mythology was Cronus, the father of Zeus (the counterpart of the Roman god Jupiter). The planet Jupiter is Saturn's nearest neighbor and the closest to it in size and composition. Like Jupiter, Saturn is a giant world formed mainly of hydrogen with no solid surface. It has a massive atmosphere, or surrounding layer of gases, with complex weather patterns.

The planet's extensive system of icy moons includes nine major moons and dozens of small ones. Some of the moons help create the rings and maintain their shape. Titan, the largest of Saturn's moons, is bigger than the planet Mercury. It is the only moon in the solar system known to have a dense atmosphere.

Four unmanned spacecraft have visited Saturn, obtaining images and data that have greatly increased knowledge about the planet. The first three—Pioneer 11 and Voyagers 1 and 2—were flybys in the late 1970s and early 1980s. The Cassini-Huygens mission arrived in 2004 for a longer study of the planet and its moons and rings.

Basic Planetary Data

Saturn's orbit lies between those of Jupiter and Uranus. Like the other outer planets—Jupiter, Uranus, and Neptune—it is made mostly of liquid and gaseous hydrogen and helium. The outer planets are significantly less dense than the rocky inner planets, such as Earth. They are also much bigger.

Size, mass, and density. Saturn is the solar system's second largest and second most massive planet, after Jupiter. The diameter at its equator is about 74,898 miles (120,536 kilometers). Since the planet has no solid surface, its diameter is measured at a level where the atmospheric pressure is 1 bar, which is equal to the pressure at sea level on Earth. Saturn's diameter is more than nine times larger than Earth's. The planet is about 95 times as massive as Earth and has more than 750 times its volume. By comparison, Jupiter is about 1.2 times as big as Saturn and 3.2 times as massive.

Saturn has the lowest mean density of any of the planets. With only about 70 percent the density of water on average, the planet would float if it could be placed in water. Earth's density, on the other hand, is about 550 percent that of water, and Jupiter's is about 130 percent.

Orbit and spin. Saturn revolves around the sun in a slightly elliptical, or oval-shaped, orbit at a mean distance of about 887 million miles (1.427 billion kilometers). Its orbit is about 9.5 times farther out than Earth's. The closest Earth and Saturn ever get to each other is about 746 million miles (1.2 billion kilometers). It takes Saturn nearly 30 Earth years to complete one revolution around the sun, so a year on Saturn is nearly 30 times longer than a year on Earth.

Since Saturn is not solid, it has no single rotation rate. However, all parts of the planet spin quickly. Clouds in the atmosphere near the equator swirl around fastest, taking about 10 hours, 10 minutes for

Facts About Saturn

Average Distance from Sun. 887,000,000 miles (1,427,000,000 kilometers).

Diameter at Equator.* 74,898 miles (120,536 kilometers).

Average Orbital Velocity. 6.0 miles/second (9.7 kilometers/second).

Year on Saturn.† 29.4 Earth years.

Rotation Period.‡ 10.8 Earth hours (as measured by Cassini spacecraft).

Day on Saturn (Solar Day). 10.8 Earth hours (as measured by Cassini spacecraft).

Tilt (Inclination of Equator Relative to Orbital Plane). 26.7°.

Atmospheric Composition. Mostly hydrogen, with some helium, small amounts of methane and ammonia, and trace amounts of other gases.

General Composition. Hydrogen, helium.

Weight of Person Who Weighs 100 pounds (45 kilograms) on Earth.* 91.4 pounds (41.5 kilograms) at equator; 123.9 pounds (56.2 kilograms) at poles (because Saturn's gravity is greater at poles).

Number of Known Moons. 60.

*Calculated at the altitude at which 1 bar of atmospheric pressure (the pressure of Earth's atmosphere at sea level) is exerted.

†Sidereal revolution period, or the time it takes the planet to revolve around the sun once, relative to the fixed (distant) stars.

‡Sidereal rotation period, or the time it takes the planet to rotate about its axis once, relative to the fixed (distant) stars.

each rotation. It takes the planet's deep interior roughly 30–40 minutes longer to complete each rotation. A day on a planet is defined by its rotation period, so a day on Saturn is about 10.8 Earth hours, or less than half as long as a day on Earth.

As in other planets, the force of the rotation causes some bulging at the equator and flattening at the poles. Saturn's rapid rotation and low average density make it the least spherical of all the planets. The diameter at its poles is about 10 percent smaller than the diameter at its equator. Jupiter actually spins a bit faster than Saturn, but its shape is less distorted. Jupiter's greater density helps it better resist the force produced by the rapid rotation.

Saturn's rotational axis is tilted about 26.7 degrees relative to the ecliptic, which is an imaginary plane passing through the sun and Earth's orbit. As Saturn orbits the sun, first one hemisphere and then the other is tipped closer to the sun. As a result, Saturn experiences seasons like Earth, which is tilted about 23.5 degrees on its axis. Since each trip around the sun takes longer for Saturn than Earth, its seasons are longer. Each season on Saturn lasts more than seven years.

The tilt of Saturn's axis also displays the rings at different angles to observers on Earth. The rings are thin and flat and always lie in the same plane as the planet's equator. As Earth and Saturn travel around the sun, Saturn and its rings are more or less tilted toward observers on Earth. At most, they are tilted about 30 degrees. The view varies over about a 30-year period, the time it takes Saturn to complete one orbit. Viewers on Earth see the sunlit northern side of the rings for about 15 years, and then the sunlit southern side for about the next 15 years. The rings are practically invisible when their thin edge is pointed directly at Earth, which happens for short periods when Earth crosses the plane of the rings.

Atmosphere

Saturn has an enormous atmosphere. It is made mostly of hydrogen, with some helium and smaller amounts of methane, ammonia, and other gases. Scientists think that small amounts of water vapor and hydrogen sulfide are probably found at lower levels of the atmosphere.

Saturn's composition, like Jupiter's, is very similar to that of the sun and other stars. Data from the Voyager mission suggest that Saturn's atmosphere is about 91 percent hydrogen and 6 percent helium by mass, making it the most hydrogen-rich atmosphere in the solar system. In comparison, hydrogen makes up about 86 percent of Jupiter's atmosphere and about 71 percent of the sun. Saturn's overall composition may be more similar to that of Jupiter and the sun, but more of Saturn's helium may have settled into its interior. Also, some research suggests that the Voyager analysis underestimated the percentage of helium and overestimated the hydrogen.

In spacecraft images of Saturn, the surface one sees is mainly clouds. Its hazy appearance is due to the appreciable atmosphere above the clouds. The highest deck of clouds is made of crystals of frozen ammonia. Farther down there are thought to be clouds made of frozen crystals of ammonium hydrosulfide and, at deeper levels, clouds of water ice crystals and ammonia droplets. All these chemicals are colorless when pure. However, the planet's clouds usually appear golden yellow-brown, perhaps because they also contain phosphorous compounds or some other chemical impurity.

Saturn and its rings are tilted at varying angles toward Earth as the two planets orbit the sun. Five images taken at about one-year intervals by the Hubble Space Telescope show the change in ring orientation from 1996 to 2000. In the first image, at bottom, the rings appear nearly edge-on, while in the last image, at top, they are opened to nearly their widest angle as seen from Earth's vicinity.

NASA and The Hubble Heritage Team (STScI/AURA) Acknowledgment: R. G. French (Wellesley College), J. Cuzzi (NASA/Ames), L. Dones (SwRI), and J. Lissauer (NASA/Ames)

In images captured by the Cassini spacecraft in the early 2000s, the atmosphere appeared blue in the northern hemisphere and yellow-brown in the south. The blue region seemed to be relatively free of the yellow-brown clouds at the highest levels. The cloudless parts of Saturn's sky were likely blue for the same reason that Earth's sky is blue—molecules of gases in the atmosphere scattering sunlight in a way favoring shorter, bluer wavelengths. Scientists are not certain why the higher clouds appeared so much thinner in the north, but it was probably a seasonal effect. As spring approached in the northern hemisphere in 2008, the atmosphere there began to turn less blue and more golden colored.

Like Jupiter, Saturn has alternating brighter and darker bands of clouds being pushed by east-west winds. Both planets also have swirling storm systems that appear as red, white, and brown ovals. The bands and ovals are less distinctive on Saturn, with subtler color differences. Saturn's atmosphere is less turbulent than Jupiter's. Occasionally, a very large storm erupts. These storms seem to occur at about 30-year intervals, or about once each orbit, suggesting they may be seasonal features. There are also two huge cyclones apparently fixed in place, one over each pole.

The bands of winds blowing eastward, in the direction of the planet's rotation, are vigorous. They alternate with bands of winds that are barely moving westward. The strongest winds blow eastward in a band from 20° N. to 20° S. Maximum wind speeds there reach nearly 1,100 miles (1,800 kilometers) per hour. Jupiter has a similar jet of winds near its equator, but Saturn's is twice as wide and its winds blow four times as fast. The fastest winds on Earth occur in tropical cyclones, or hurricanes, and are much slower. Only in extreme cases do they reach sustained speeds of more than 150 miles (240 kilometers) per hour.

Saturn is a stormy world. A giant thunderstorm called the Dragon Storm appears as an orange feature above and to the right of center in a false-color, near-infrared composite of images (left) taken by the Cassini spacecraft. Scientists believe it is a long-lived storm deep in the atmosphere that flares up from time to time. A huge hurricane-like storm whirls over Saturn's south pole in another Cassini image (right). The storm is about two thirds as big across as Earth. Unlike earthly hurricanes, this storm remains in place and is not fueled by an ocean below.

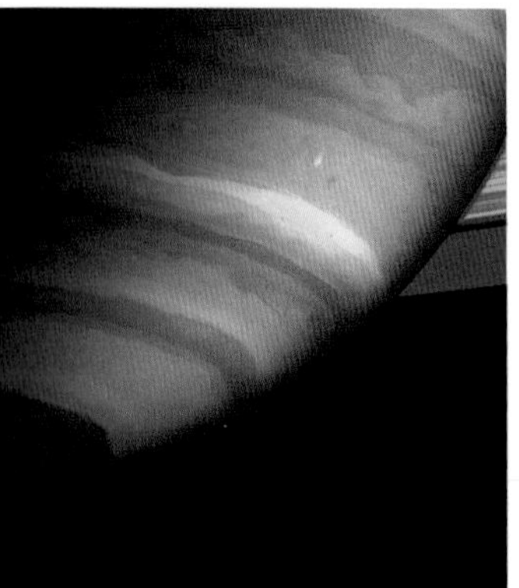

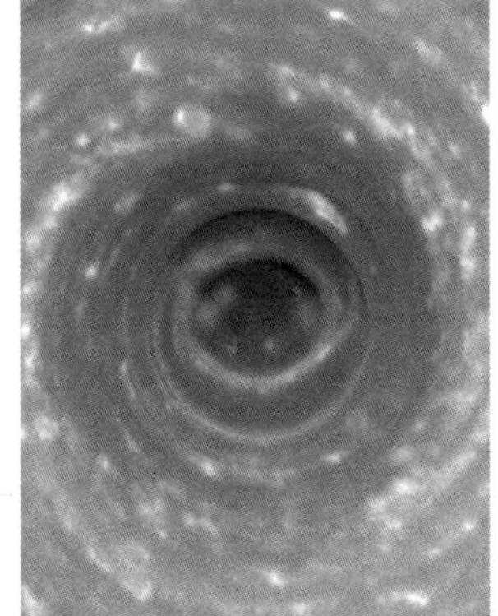

(Left and right) NASA/JPL/Space Science Institute

Saturn's atmosphere

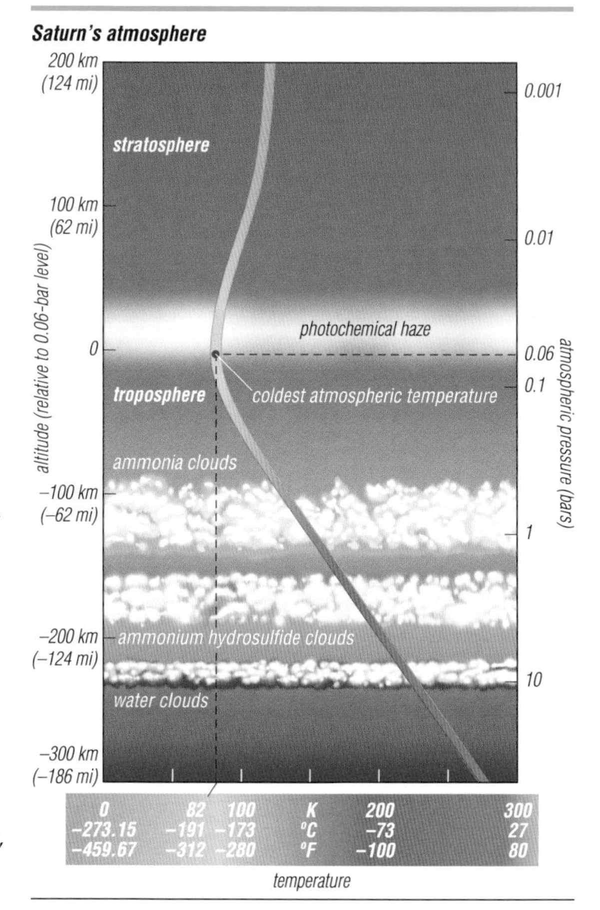

A graph shows the temperatures and pressures at different levels of Saturn's atmosphere. In the layer called the stratosphere, it gets colder with increased pressure. The bottom of this layer has Saturn's coldest temperature—82 Kelvin (K), which is about –312° F (–191° C). The pressure at that level is 0.06 bar, or six hundredths of the pressure at sea level on Earth. In the troposphere below, it get warmer with increased pressure.

Saturn's alternating bands of winds are remarkably symmetrical. Each band north of the equator usually has a counterpart south of the equator with about the same width and wind velocity. This suggests to scientists that the pairs of bands may be connected in some way deeper in the atmosphere.

The pressure increases with depth in Saturn's atmosphere. At the lower levels, where the pressure is very high, the hydrogen is probably crushed into a liquid. The temperature in the atmosphere also varies; at its coldest, it is about –312° F (–191° C). As on Earth, the temperature gets colder with altitude in the lowest level of the atmosphere but hotter with altitude in a middle level. In the highest level, the temperature is fairly constant.

Interior

The temperatures and pressures in Saturn's interior are very high, and they increase with depth. As in Jupiter, the interior consists largely of hydrogen, which the immense pressure squeezes into a liquid. About halfway down between Saturn's cloud tops and center, the temperature is probably about 10,300° F (5,730° C). The pressure is thought to be some 2 million times greater than at sea level on Earth. Under those conditions, the hydrogen is probably compressed into a liquid metallic state. In this state, the electrons are stripped away from the atomic nuclei, so the hydrogen would conduct electricity like a metal.

Saturn's central, liquid metallic region is denser than Jupiter's. Scientists have determined this by analyzing Saturn's gravity field. The planet's gravity is stronger at the poles than at the equator. The distortions in its gravity field are directly related to the relative amount of mass concentrated in its interior rather than in its atmosphere. This analysis suggests that Saturn's central regions are about half hydrogen by mass and half denser materials. Jupiter's central regions are thought to contain about two thirds hydrogen and only one third denser matter. Some of the denser material must be helium, which may be more concentrated in Saturn's interior than Jupiter's. Saturn's dense core is likely a mixture of rock and ice with about 10 to 20 times the mass of Earth.

Like Jupiter, Saturn radiates nearly twice as much energy as it receives from the sun, mostly as heat. This means that the planet must generate some of its own heat. Much of this energy is probably left over from when the planet formed some 4.6 billion years ago. Since then, the planet has slowly cooled down, gradually emitting heat. Scientists believe that some of the heat that Saturn produces probably comes from helium settling into its interior. It is thought that the helium separates out of the hydrogen and forms droplets, which sink toward the center. The friction of the droplets pushing against the other matter would create heat. This process is also thought to occur in Jupiter, but to a much lesser degree.

Magnetic Field and Magnetosphere

Saturn's magnetic field is much stronger than Earth's but much weaker than Jupiter's. It has two poles, north and south, like a giant bar magnet. As on Jupiter, the orientation of the poles is opposite that currently found on Earth. This means that a compass on Saturn or Jupiter would point south.

The planet's magnetic field dominates a large region of space called its magnetosphere. This region is shaped like a teardrop. The rounded part extends about 750,000 miles (1,200,000 kilometers) from the planet on the side facing the sun. There, the magnetosphere holds off the solar wind, a flow of electrically charged particles from the sun. On the side opposite the sun, the solar wind pulls the magnetosphere out into a very long tail.

The inner part of Saturn's magnetosphere traps clouds of highly energetic protons and other electrically charged particles. The clouds travel around the planet in large regions called belts. These belts are similar to the radiation belts of Earth (called the Van Allen belts) and Jupiter. When charged particles collide with the hydrogen in the atmosphere above Saturn's poles, it causes glowing auroras, like Earth's northern and southern lights. Earth's auroras typically last only minutes. Saturn's, however, can last for days.

Ring System

Saturn's spectacular rings have long been admired for their beauty. Its prominent rings are brighter and broader than the faint, narrow principal rings of the other outer planets. They are the easiest rings to see from Earth and so were the first to be discovered. Galileo observed them through an early telescope in 1610, but he did not identify them as rings. In 1655, using a more powerful telescope, the scientist Christiaan Huygens was able to see a flat, apparently solid ring around Saturn. The scientist James Clerk Maxwell demonstrated mathematically in 1857 that the rings could not be solid but must be composed of many small particles. This theory was confirmed by observations made by James Keeler in the 1890s. In the 1980s the cameras of the Voyager spacecraft revealed that there are really hundreds of thousands of individual rings (or "ringlets") around Saturn. The Cassini spacecraft discovered additional rings and structures within them in the 2000s.

The main rings have a diameter of about 170,000 miles (270,000 kilometers), and the fainter outer rings extend much farther. The entire ring system spans some 600,000 miles (1,000,000 kilometers). The rings are very thin, however, reaching a maximum thickness of roughly 300 feet (100 meters). They are made of countless particles, largely of water ice and dust, all orbiting Saturn like tiny moons. The particles range in size from no bigger than a speck of dust to the size of cars or houses. There are many more small particles than large ones. The individual particles that make up

A dramatic backlit view of Saturn reveals its ring system as never before seen. From the vicinity of Earth, the planet and its rings always appear nearly fully illuminated. This backlit view is a mosaic of many images captured by the Cassini spacecraft while flying in Saturn's shadow, with the sun on the opposite side of the planet. The colors have been exaggerated to bring out greater detail. From the planet's night side, the rings appear bright, and light reflected off the rings partly illuminates Saturn itself. Easily visible surrounding the stunning main rings are the narrow G ring and the diffuse E ring, which is the outermost known ring. Scientists also discovered two new faint rings from the data collected by Cassini while in the planet's shadow.

NASA/JPL/Space Science Institute

the rings have not been seen directly. However, scientists can determine their size distribution and composition by the way different parts of the rings reflect light, radio signals, and other radiation.

The rings occur in groups. The three main groups, from farthest to closest to Saturn, are called the A ring, the B ring, and the C ring. These three rings are visible from Earth through a telescope. The B ring is the broadest, thickest, and brightest ring. Other rings with lower densities of particles lie outside the main groups. The D ring lies between the C ring and Saturn, and the F, G, and E rings extend out from the A ring. Many gaps occur between the rings. The gaps are regions where far fewer particles are found. Some of the major gaps are named after astronomers who studied Saturn. The largest gap, between the A and B rings, is called the Cassini division. It was named after the astronomer Gian Domenico Cassini, who discovered the gap in 1675.

The rings have a complex structure, and they interact with many of the planet's moons in a number of ways. The outermost rings seem to consist of small particles that are continually shed by moons. The Cassini spacecraft revealed that erupting geysers of water vapor and water ice on the moon Enceladus feed particles into the E ring. In addition, moons send particles into the outer rings when objects collide with them. In fact, Saturn's main rings may have been formed by an icy moon or moons that completely broke apart, perhaps tens of millions of years ago.

Several small inner moons orbit Saturn embedded within the ring system. As these moons and others orbit the planet, their gravity affects the distribution of the ring particles. For instance, the moon Pan acts as a "sweeper," clearing particles from its orbital vicinity. This creates a gap in the rings called the Encke gap. Other moons act as "shepherds," by helping to keep the ring material in place. The shepherd moons Pandora and Prometheus orbit on either side of the F ring and constrain its particles into a narrow band. Their gravity is thought to be responsible for that ring's braided and knotted appearance.

Moons

At least 60 moons orbit Saturn, in addition to the chunks of material in the rings. Nine of them have diameters greater than 125 miles (200 kilometers). In order of distance from Saturn, these major moons are Mimas, Enceladus, Tethys, Dione, Rhea, Titan, Hyperion, Iapetus, and Phoebe. All of them were discovered before the 20th century. The rest of Saturn's moons were discovered in the late 20th or early 21st centuries, in images captured by spacecraft or by powerful Earth-based telescopes equipped with very sensitive electronic detectors. Many of those moons are quite small.

Saturn's inner moons are called regular because they orbit in nearly circular paths in or near the plane of Saturn's orbit. The inner moons include all of Saturn's major moons except Phoebe, plus about a dozen others. The planet's outer moons are irregular, having highly elongated or tilted orbits, or both. The eight large inner moons are thought to have formed along with Saturn about 4.6 billion years ago. The outer moons probably started out as other objects but came too close to Saturn and were captured by its gravity.

Saturn's three main rings appear in a natural-color composite of six images taken by the Cassini spacecraft. The view is from below the ring plane, with the rings tilted at an angle of about 4°. The major gaps in the rings are labeled.

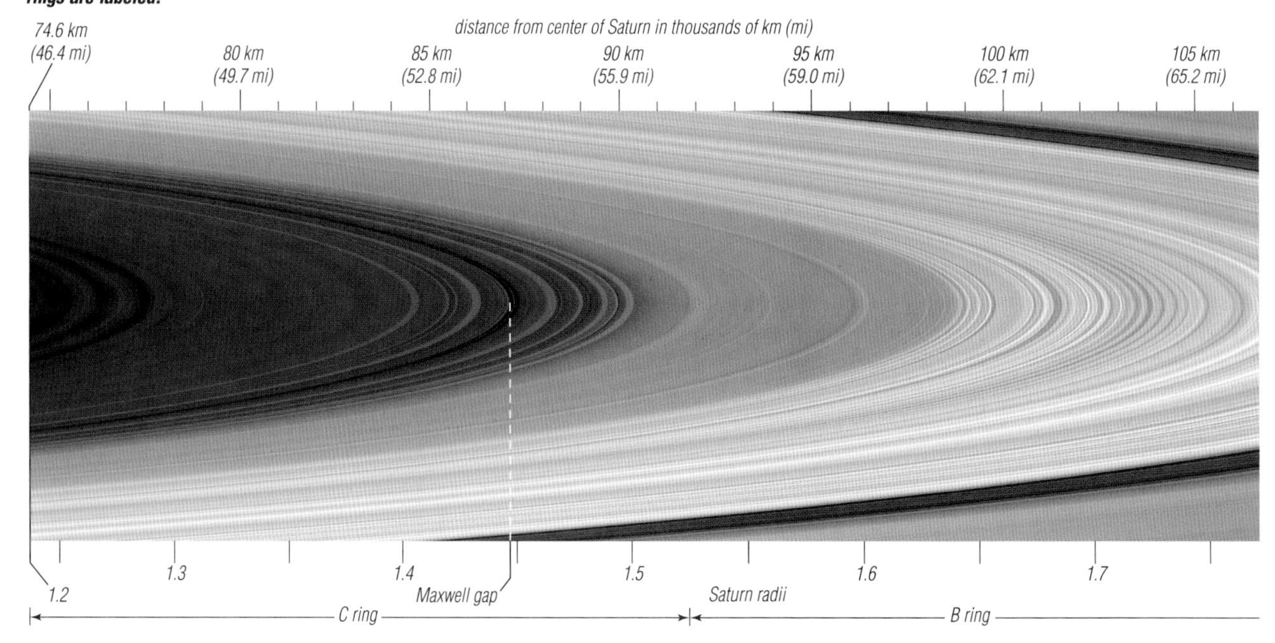

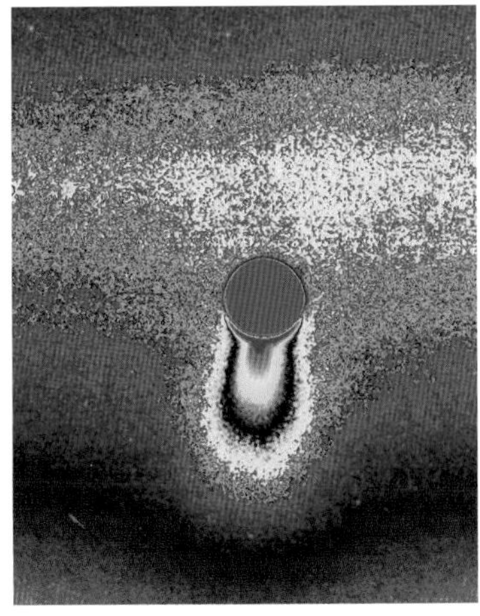

(Left) ESA/NASA/JPL/University of Arizona; (center and right) NASA/JPL/Space Science Institute

Saturn's moons are fascinating worlds. Frozen rocks on the frigid surface of Titan appear in an image (left) captured by the Huygens probe. The Cassini orbiter photographed Dione (center) above the planet's thin rings, seen edge-on. At back is Saturn itself, marked by the rings' shadows. Jets of water vapor and ice erupt from Enceladus (right) in a color-coded Cassini image. The jets feed ice particles into Saturn's E ring, which is the bright band above and behind the moon.

The planet's largest moon, Titan, is remarkable in several ways. Its diameter is some 3,200 miles (5,150 kilometers), making it the second largest moon in the solar system. Only Jupiter's moon Ganymede is bigger. Unique among all moons, Titan has clouds and a very dense atmosphere. The atmosphere is even denser than Earth's, with a surface pressure that is about 1.5 times greater. Like Earth, Titan has an atmosphere that is mostly nitrogen. Titan's atmosphere also has about 5 percent methane and trace amounts of other gases. Methane might play a role in its atmosphere similar to that of water vapor on Earth. Liquid methane probably rains out of the clouds.

Titan is thought to be about half rock and half ices (mostly water ice mixed with some frozen ammonia and methane). Just above the surface it is only about −290° F (−179° C). A thick, orangish haze envelops the moon, so little was known about its surface until the Cassini-Huygens mission arrived. It discovered that the surface is fairly young and is sculpted by the wind. What seem to be vast sand dunes appear in radar images of areas near the equator. Scientists expected to find oceans of liquid methane, but none were found. However, radar images of the polar regions show what are probably hundreds of lake beds—some dry and some apparently filled with liquid, likely methane or

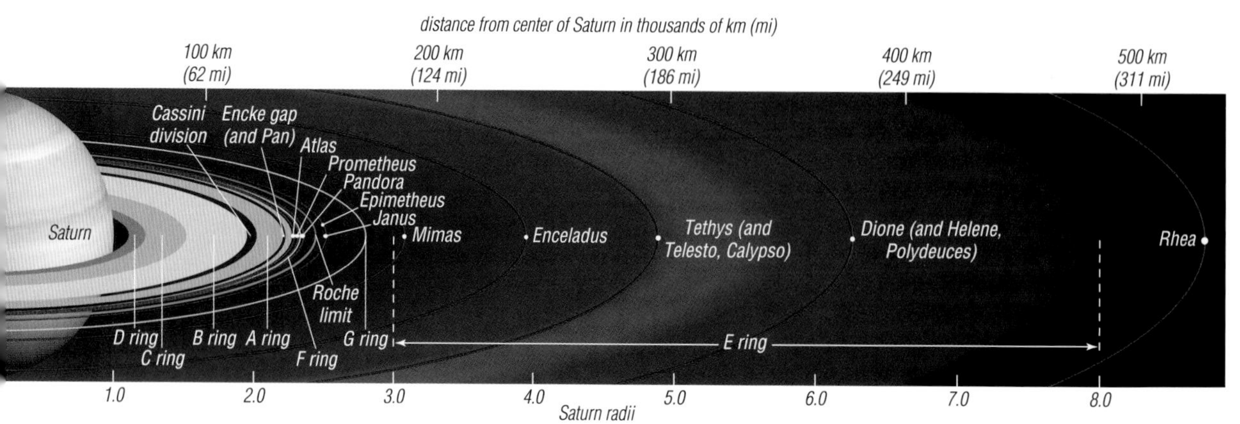

Many of Saturn's moons orbit the planet from within the extensive ring system. A diagram shows the full ring system, with red lines indicating the orbits of some of the associated moons.

ethane. Dark channels, perhaps carved by methane rain, are common on the surface.

The planet's other major moons are much smaller. They have low average densities and mostly bright, reflective surfaces that are rich in ices, mostly water ice. Their surfaces are so cold that the ice behaves like rock and can retain craters, which are the scars of collisions with other objects. In general, the more craters a solid body has, the older its surface. Mimas, Tethys, Dione, Rhea, Hyperion, Iapetus, and Phoebe are all heavily cratered. Mimas has one crater that is about a third as big across as Mimas itself. It is one of the largest-known craters in the solar system in relation to the size of the object. Though largely crater-

Major Moons of Saturn

	Mimas	**Enceladus**	**Tethys**
average distance from center of planet	115,275 mi (185,520 km)	147,900 mi (238,020 km)	183,095 mi (294,660 km)
diameter or dimensions	250 mi (400 km)	310 mi (500 km)	660 mi (1,060 km)
density	1.1 grams per cubic centimeter	1.3 grams per cubic centimeter	1.0 grams per cubic centimeter
orbital period	0.95 Earth days	1.4 Earth days	1.9 Earth days
rotation period	synchronous (same as orbital period)	synchronous (same as orbital period)	synchronous (same as orbital period)
discovery year, discoverer	1789, William Herschel	1789, William Herschel	1684, Gian Domenico Cassini
	Dione	**Rhea**	**Titan**
average distance from center of planet	234,505 mi (377,400 km)	327,490 mi (527,040 km)	759,220 mi (1,221,850 km)
diameter or dimensions	695 mi (1,120 km)	950 mi (1,530 km)	3,200 mi (5,150 km)
density	1.5 grams per cubic centimeter	1.3 grams per cubic centimeter	1.9 grams per cubic centimeter
orbital period	2.7 Earth days	4.5 Earth days	15.9 Earth days
rotation period	synchronous (same as orbital period)	synchronous (same as orbital period)	synchronous (same as orbital period)
discovery year, discoverer	1684, Gian Domenico Cassini	1672, Gian Domenico Cassini	1655, Christiaan Huygens
	Hyperion	**Iapetus**	**Phoebe**
average distance from center of planet	920,310 mi (1,481,100 km)	2,212,890 mi (3,561,300 km)	8,048,000 mi (12,952,000 km)
diameter or dimensions	230 x 175 x 140 mi (370 x 280 x 225 km)	890 mi (1,435 km)	140 mi (220 km)
density	0.5 grams per cubic centimeter	1.0 grams per cubic centimeter	1.6 grams per cubic centimeter
orbital period	21.3 Earth days	79.3 Earth days	550.4 Earth days (retrograde)
rotation period	chaotic (no regular rotation period)	synchronous (same as orbital period)	0.4 Earth day
discovery year, discoverer	1848, William Bond, George Bond, William Lassell	1671, Gian Domenico Cassini	1899, William Henry Pickering

scarred, Dione and Rhea also have smoother plains and other features that suggest their interiors may have been geologically active more recently. Parts of their icy surfaces seem to have melted and refrozen at some point.

Portions of Enceladus are geologically active today. It is Saturn's brightest moon, with a surface of almost pure water ice. A hot spot near its south pole fuels geysers that spew large amounts of water vapor and water ice. The water ice particles from these eruptions form Saturn's E ring. The surface of Enceladus has few large craters overall and some crater-free areas that must have formed fairly recently. This suggests that it may have been internally active recently in other areas besides the south.

Hyperion is an unusual moon. It is the only known moon in the solar system that does not rotate regularly. Instead, it tumbles along in an apparently random fashion along its orbit. It is also the largest known moon with an irregular shape. Its very low density—only about half that of water—suggests that its interior may be full of holes.

Iapetus is remarkable for having one very dark half and one bright half, for reasons that are not yet known. Its surface has the greatest variation in brightness found in the solar system. The bright side is likely water ice. The dark side may be coated in a layer that is rich in complex organic molecules.

Spacecraft Exploration

The first spacecraft to encounter Saturn was Pioneer 11. It was launched by the United States National Aeronautics and Space Administration (NASA) in 1973. After completing its original mission at Jupiter, the craft was reprogrammed and sent to Saturn. Pioneer flew within about 13,000 miles (21,000 kilometers) of Saturn's cloud tops in 1979. It transmitted data and close-up photographs that enabled scientists to identify previously unknown moons and the F ring.

NASA's Voyager 1 and 2 spacecraft, launched in 1977, were outfitted with more sophisticated equipment. After surveying Jupiter, the two spacecraft reached Saturn in 1980–81. The Voyagers returned tens of thousands of images. The structure of the rings was found to be far more complex than could be seen in the lower-resolution Pioneer images or with the best telescopes on Earth. The craft also photographed previously unknown shepherd moons among the rings.

The Cassini-Huygens mission was launched in 1997 as a joint venture between NASA, the European Space Agency (ESA), and the Italian Space Agency. The Cassini spacecraft began orbiting Saturn in 2004 in order to study the planet, its moons, and its rings for several years. The orbiter carried on board the ESA-built Huygens probe, which it released toward Titan. The probe parachuted through Titan's atmosphere and landed on its surface in early 2005, providing the first look at the surface of the haze-shrouded moon. It transmitted data and photographs for about three hours during the descent and on the surface. It was the first craft to land on a moon other than Earth's.

Exploring Saturn: Major Spacecraft Missions

Spacecraft	Type of Mission	Launch Date	Country or Agency	Highlights
Pioneer 11	flyby	April 5, 1973	U.S.	First craft to visit Saturn. After flying past Jupiter, flew past Saturn on Sept. 1, 1979. Revealed radiation belts in magnetosphere and a previously unknown ring.
Voyager 2	flyby	Aug. 20, 1977	U.S.	First craft to visit all four outer planets. After visiting Jupiter, flew past Saturn on Aug. 25, 1981, then sent on path to take it out of solar system. Along with Voyager 1, took photographs and measurements that revealed many previously unknown details of Saturn, including spokes, braids, and kinks in ring system, and composition of Titan's atmosphere.
Voyager 1	flyby	Sept. 5, 1977	U.S.	Twin of Voyager 2. After visiting Jupiter, flew past Saturn on Nov. 12, 1980, then sent on path to take it out of solar system.
Cassini-Huygens	orbiter and lander	Oct. 15, 1997	U.S., ESA, Italy	Cassini craft began orbiting Saturn on June 30, 2004. Released Huygens probe, which landed on Titan on Jan. 14, 2005, becoming first craft to land on solid body in the outer solar system. Mission returned wealth of data and images. Among its discoveries were a new radiation belt, a slightly longer rotation period for Saturn than previously measured, new moons, complex interactions between moons and rings, ice geysers on Enceladus, and methane rain on Titan.

Artist's rendering of Cassini and Huygens at Titan courtesy of NASA/JPL

Jet Propulsion Laboratory/National Aeronautics and Space Administration
In colors visible to the unaided eye, Uranus appears as a nearly featureless blue-green sphere. The view shows the planet's southern hemisphere. It was produced from images taken by the Voyager 2 spacecraft.

URANUS

The seventh planet from the sun is Uranus. It is one of the giant outer planets with no solid surfaces. Although Uranus is not as big as Jupiter or Saturn, more than 60 Earths would fit inside it. The planet is most similar in size and composition to Neptune, its outer neighbor. Like Neptune, Uranus is blue-green because of the small amount of methane in its atmosphere, or surrounding layer of gases.

Uranus was not known to ancient astronomers. At its brightest, the planet is just barely visible to the unaided eye. It was seen in early telescopes several times but was thought to be just another star. In 1781, as part of a telescopic survey of the stars, English astronomer William Herschel discovered "a curious either nebulous star or perhaps a comet." This unusual object soon proved to be a planet, the first to be identified in modern times. It was later named Uranus, the personification of the heavens in ancient Greek mythology. Even with the more powerful telescopes available in the 21st century, the distant planet is difficult to observe in great detail. Much of what is known about it comes from close-up photographs and measurements taken in 1986 by Voyager 2, the only spacecraft to visit Uranus.

Basic Planetary Data

Uranus orbits the sun between the orbits of Saturn and Neptune. Those three planets plus Jupiter are the four outer planets called Jovian, or Jupiter-like. Composed mainly of liquids and gases, they are much less dense than the four inner planets. Like the other outer planets, Uranus has a system of rings and several moons.

Size, mass, and density. Uranus is the third largest planet in the solar system, after Jupiter and Saturn. Uranus is about four times bigger than Earth. Its diameter at the equator is about 31,763 miles (51,118 kilometers), as measured at the level of the atmosphere where the pressure is the same as at sea level on Earth. The planet is slightly larger than Neptune, but Neptune is about 1.2 times more massive. Uranus' density is quite low—only about 1.3 times that of water, compared with 1.6 for Neptune and 5.5 for rocky Earth.

Orbit and spin. Like all the planets, Uranus orbits the sun in a slightly elliptical, or oval-shaped, orbit. With an average distance from the sun of about 1.78 billion miles (2.87 billion kilometers), Uranus is about 19 times farther from the sun than Earth is. The closest the planet ever gets to Earth is some 1.7 billion miles (2.7 billion kilometers) away. It takes Uranus about 84 Earth years to complete just one trip around the sun. This means that a year on Uranus is about 84 times as long as a year on Earth.

A day on Uranus, however, is shorter than one on Earth. Uranus completes one rotation on its axis in about 17 Earth hours, compared with about 24 hours for Earth. This rapid rotation causes its polar regions to flatten slightly and its equator to bulge. The diameter at its poles is about 2 percent smaller than that at its equator.

Facts About Uranus

Average Distance from Sun. 1,783,950,000 miles (2,870,990,000 kilometers).

Diameter at Equator.* 31,763 miles (51,118 kilometers).

Average Orbital Velocity. 4.2 miles/second (6.8 kilometers/second).

Year on Uranus.† 84.0 Earth years.

Rotation Period.‡ 17.2 Earth hours (retrograde).

Day on Uranus (Solar Day). 17.2 Earth hours (retrograde).

Tilt (Inclination of Equator Relative to Orbital Plane). 97.9°.

Atmospheric Composition. Mostly hydrogen, with some helium, a small amount of methane, and trace amounts of other gases.

General Composition. Melted ices, silicates, and hydrogen and helium.

Weight of Person Who Weighs 100 pounds (45 kilograms) on Earth.* 88.7 pounds (40.2 kilograms).

Number of Known Moons. 27.

*Calculated at the altitude at which 1 bar of atmospheric pressure (the pressure of Earth's atmosphere at sea level) is exerted.

†Sidereal revolution period, or the time it takes the planet to revolve around the sun once, relative to the fixed (distant) stars.

‡Sidereal rotation period, or the time it takes the planet to rotate about its axis once, relative to the fixed (distant) stars.

Technically, Uranus spins on its axis in retrograde motion, or the direction opposite that of most other planets. However, it is a bit misleading to describe its rotation that way, since Uranus lies nearly on its side. Unlike in any other planet, its rotational axis is tilted an unusually large 98 degrees relative to the plane in which it orbits. Scientists think that Uranus may have been knocked into this alignment early in its history by one or more violent collisions with other bodies.

Each season on Uranus lasts about 21 Earth years. Since the planet is nearly tipped on its side, as it orbits it points first one pole toward the sun, then its equator, and then the other pole. As a result, summers and winters are extreme, with one hemisphere bathed in sunlight for many years during its summer, while the other hemisphere is plunged in constant darkness for its long winter. The sunshine is more evenly distributed during spring and fall, when the equator is pointed toward the sun. However, heat seems to be fairly evenly distributed year-round. The two hemispheres are probably always about the same temperature, probably because the atmosphere transfers and stores heat well.

Atmosphere

Like the other outer planets, Uranus has a massive atmosphere with a composition similar to that of the sun and other stars. Scientists think it is roughly three quarters hydrogen and a quarter helium by mass, plus a small amount of methane and probably trace amounts of water, ammonia, and other substances. The highest clouds are very bright and are formed of frozen methane. Farther down, there are perhaps clouds of frozen water and ammonium hydrosulfide. The lower parts of the atmosphere, in which clouds form, are quite cold, and the temperature there decreases with increased altitude. The coldest part of the atmosphere is about –366° F (–221° C). The temperature rises remarkably in the upper atmosphere, however, reaching 890° F (480° C).

Unlike Jupiter and Saturn, Uranus appears nearly featureless in visible light. Faint bands of clouds are revealed in images taken at other wavelengths of light or processed to show extreme contrast. The bands of clouds are parallel to the equator. As on Earth and Neptune, winds travel west in a zone near the equator and east in zones at higher latitudes. The winds are several times stronger than Earth's but weaker than Neptune's. The atmosphere of Uranus seems to be calmer than those of the other outer planets. Spots observed on the planet are thought to be storms, but they are smaller and fewer than those seen on Jupiter, Saturn, and Neptune.

Interior

Pressures and temperatures are very high inside the planet, so its interior must be liquid. Scientists think that Uranus is composed mainly of melted ices of water, methane, and ammonia, with some molten silicate rock and metals, and a smaller amount of hydrogen and helium. At its center the planet might have a core of rock and metal. However, scientists think that the rock and metal are more likely to be

Faint bands of clouds and individual cloud features are revealed in composite images of Uranus taken at near-infrared wavelengths with the Keck II telescope in Hawaii. The two composites show opposite hemispheres. In both images the planet's south pole is oriented to the left and tilted slightly above the horizontal. The rings appear red because of the color processing used to make the highest-level clouds (seen mostly in the northern hemisphere) appear white.

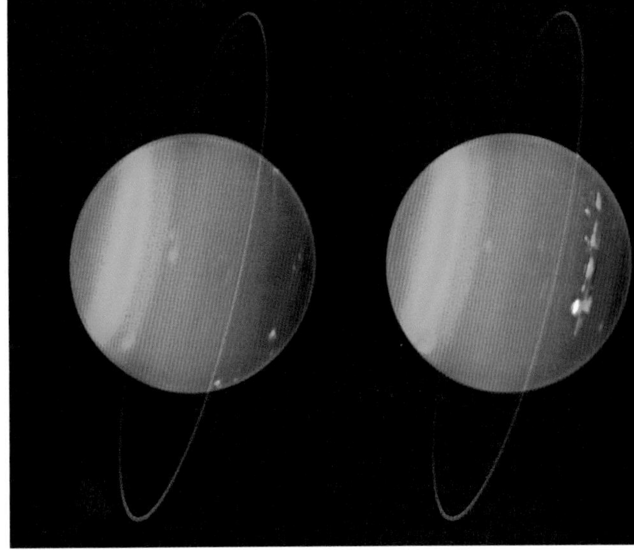

Lawrence Sromovsky, University of Wisconsin, Madison/W.M. Keck Observatory

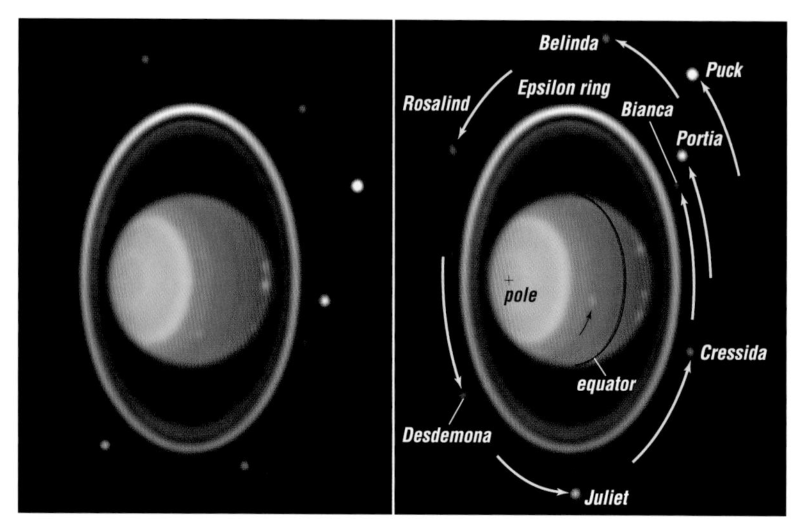

Uranus' southern hemisphere, ring system, and eight of its small inner moons appear in two false-color images made 90 minutes apart by the Hubble Space Telescope. Comparison of the images reveals the orbital motion of the moons along Uranus' equatorial plane and the counterclockwise rotation of clouds in the planet's atmosphere.
Erich Karkoschka, University of Arizona, Tuscon, and NASA

spread throughout the fluid interior than in a separate layer.

The interior of Uranus is more like that of Neptune than like the interiors of Jupiter and Saturn, which are mostly hydrogen and helium. As in Neptune, melted ices, rock, and metal make up a much greater part of the mass. For some unknown reason, Uranus does not seem to generate as much internal heat as the three other outer planets. Those planets radiate almost twice as much heat as they receive from the sun, but Uranus emits just a bit more heat than it receives.

Like most other planets in the solar system, Uranus produces its own magnetic field. It is similar to a bar magnet, with a north pole and a south pole. As on Earth, a compass would point north. However, Uranus' magnetic north pole is tilted an exceptionally great 58.6 degrees from its rotational north pole, compared with an 11.5-degree inclination for Earth. Only Neptune's magnetic field is similarly tilted.

Ring System

Uranus has a system of about a dozen narrow rings. Like Saturn's rings, they are made up of countless particles, each orbiting the planet like a small moon. The particles in Uranus' rings are much darker than those found in Saturn's bright, icy rings. Also, Saturn's rings have a much higher percentage of dust and tiny particles. Most of the objects forming Uranus' rings are larger than about 4.6 feet (1.4 meters) across. The small

Major Moons of Uranus

	Miranda	Ariel	Umbriel
average distance from center of planet	80,655 mi (129,800 km)	118,830 mi (191,240 km)	165,265 mi (265,970 km)
diameter or dimensions	290 mi (470 km)	720 mi (1,160 km)	727 mi (1,170 km)
density	1.2 grams per cubic centimeter	1.7 grams per cubic centimeter	1.4 grams per cubic centimeter
orbital period	1.4 Earth days	2.5 Earth days	4.1 Earth days
rotation period	synchronous (same as orbital period)	synchronous (same as orbital period)	synchronous (same as orbital period)
discovery year, discoverer	1948, Gerard P. Kuiper	1851, William Lassell	1851, William Lassell

	Titania	Oberon
average distance from center of planet	270,820 mi (435,840 km)	362,010 mi (582,600 km)
diameter or dimensions	981 mi (1,578 km)	946 mi (1,522 km)
density	1.7 grams per cubic centimeter	1.6 grams per cubic centimeter
orbital period	8.7 Earth days	13.5 Earth days
rotation period	synchronous (same as orbital period)	synchronous (same as orbital period)
discovery year, discoverer	1787, William Herschel	1787, William Herschel

amount of dust in its rings seems to be constantly replenished. The dust may be knocked off small moons when objects hit their surfaces.

Voyager 2 recorded two small moons, Cordelia and Ophelia, that orbit on either side of one of the rings. The gravity of the two moons confines the ring particles into a narrow band, so they are called shepherd moons. Other small moons not yet discovered may be shepherds for the other rings.

Moons

Uranus has five major moons—Miranda, Ariel, Umbriel, Titania, and Oberon—and more than 20 smaller ones. Some of the small moons orbit near the rings, while others orbit beyond the major moons. The outer small moons are irregular, meaning that they have highly elongated or tilted orbits, or both. The inner small moons and the five major moons have nearly circular orbits that lie in about the same plane as Uranus' orbit.

The planet's five major moons are probably mostly water ice and rock. The four largest are thought to be about 60 percent ice and 40 percent rock. Miranda, the smallest of the five major moons, has a lower density, so it probably has a greater percentage of ice. The surfaces of all five major moons seem to contain dirty water ice. Umbriel and Oberon have many craters, large and small, like the highlands of Earth's moon. Like the moon's, their large craters probably date back more than 4 billion years. Titania and Ariel have fewer large craters but about as many small craters. This suggests that Titania and Ariel have younger surfaces. Narrow canyons are found on all the major moons. They may have formed by the cracking of the crusts as the moons expanded.

Miranda has the largest canyons, with some being as much as 50 miles (80 kilometers) wide and 9 miles (15 kilometers) deep. Scientists think that all the water in its interior may once have been liquid. As the water froze, the moon would have expanded, causing the crust to fracture. Miranda has an odd jumble of different types of terrain. It mostly has heavily cratered, ancient surfaces. Other areas have fewer craters, sets of curving grooves, winding valleys, or steep cliffs. Miranda may have been broken apart by collisions with other objects and then reassembled to form the strange patchwork of terrains now observed.

U.S. Geological Survey/NASA/JPL

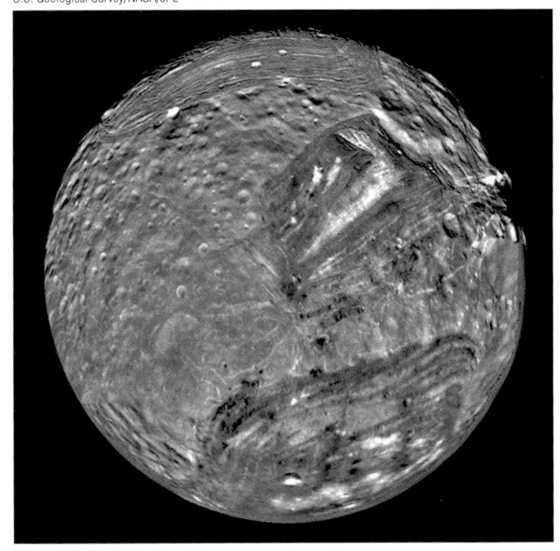

Miranda has the most diverse terrain of Uranus' moons. A mosaic of images taken by Voyager 2 shows the moon's south polar region, which has heavily cratered areas and large patches of lightly cratered regions marked with parallel bands and ridges. Such patches have not been found on any other body in the solar system.

The different terrains may have formed instead by eruptions and other internal geologic activity.

Spacecraft Exploration

In 1977 the United States launched the unmanned probe Voyager 2 on a mission to Jupiter and Saturn, with the hopes that it could later be sent to Uranus and Neptune. After the spacecraft visited the two closer giant planets, its course was indeed changed to send it to the two outer giant planets. Voyager 2 became the first—and so far only—spacecraft to encounter Uranus, in 1986, and Neptune, in 1989.

After passing through Uranus' ring system, Voyager 2 flew to within about 66,500 miles (107,000 kilometers) of the planet's center. It measured the size and mass of Uranus and its major moons, detected and measured the magnetic field, and determined the rotation rate of the planet's interior. Its close-up photographs uncovered weather patterns in the atmosphere of Uranus and the surface conditions of the major moons; they also revealed for the first time many smaller moons.

Exploring Uranus: Spacecraft Missions

Spacecraft	Type of Mission	Launch Date	Country or Agency	Highlights
Voyager 2	flyby	Aug. 20, 1977	U.S.	First and only spacecraft to visit Uranus. Took advantage of rare planetary alignment to fly by all four outer planets. After visiting Jupiter and Saturn, flew past Uranus on Jan. 24, 1986, before visiting Neptune and then being sent on path to take it out of solar system. At Uranus, collected data and took about 8,000 photographs, including the first close-up images of the planet, rings, and moons.

NASA/JPL

Clouds appear in Neptune's dynamic atmosphere in an image captured by Voyager 2 in 1989. At the center is the Great Dark Spot, a swirling storm system the size of Earth, and its associated methane-ice clouds. The giant storm system disappeared by 1991.

NEPTUNE

The eighth and farthest planet from the sun is Neptune. It is always more than 2.5 billion miles (4 billion kilometers) from Earth, making it too far to be seen with the unaided eye. It was the second planet, after Uranus, to be discovered through a telescope but the first planet to be found by people specifically searching for one. In the mid-1800s several astronomers began looking for a planet beyond Uranus, in part because Uranus did not move along its orbit exactly as expected. Scientists thought that these slight differences could be caused by the gravitational pull of another planet, and they were right. Several people can be credited with Neptune's discovery. John Couch Adams and Urbain-Jean-Joseph Le Verrier independently calculated the planet's probable location, while in 1846 Johann Gottfried Galle and his assistant Heinrich Louis d'Arrest were the first to identify it in the night sky. The new planet was named Neptune after the ancient Roman god of the sea.

Relatively little was known about the distant planet until the Voyager 2 spacecraft—the only mission to Neptune—flew by it in 1989. The planet that Voyager uncovered is a stormy, windswept world with a vivid blue hue. Its highly active atmosphere is surprising, since it receives so little sunlight to power its weather systems. Like the other giant outer planets, Neptune has no solid surface. It also has a system of rings and more than a dozen moons.

Basic Planetary Data

Neptune's orbit lies beyond that of Uranus, the planet it most resembles in size and composition. It is one of the four giant outer planets, along with Jupiter, Saturn, and Uranus. The four outer planets are much larger than the four inner planets. They are also much less dense, being composed mainly of liquids and gases.

Size, mass, and density. Neptune is the smallest of the four giant outer planets. The diameter at its equator is about 30,775 miles (49,528 kilometers), as measured at a level of the atmosphere where the pressure is 1 bar (the pressure at sea level on Earth). This makes it slightly smaller than Uranus but nearly four times as big as Earth. Neptune's mass is about 1.2 times greater than Uranus', however, and more than 17 times greater than Earth's. It is the third most massive planet in the solar system, after Jupiter and Saturn. Like the other outer planets, Neptune has a low density—only about 1.6 times the density of water. However, it is the densest of the four, being roughly 25 percent denser than Uranus.

Orbit and spin. Neptune revolves around the sun in a nearly circular orbit at a mean distance of about 2.8 billion miles (4.5 billion kilometers). On average, it is more than 30 times farther from the sun than Earth is. Because of its great distance from the sun, Neptune takes nearly 164 Earth years to complete one orbit.

The dwarf planet Pluto is usually farther from the sun than Neptune is. About every 248 years, however, Pluto's highly eccentric (elongated) orbit brings it inside Neptune's orbit. Neptune is then farther from the sun than Pluto is for a period of some 20 Earth years, as it last was in 1979–99. Pluto was classified as

Facts About Neptune

Average Distance from Sun. 2,795,083,000 miles (4,498,250,000 kilometers).

Diameter at Equator.* 30,775 miles (49,528 kilometers).

Average Orbital Velocity. 3.4 miles/second (5.5 kilometers/second).

Year on Neptune.† 163.7 Earth years.

Rotation Period.‡. 16.1 Earth hours.

Day on Neptune (Solar Day). 16.1 Earth hours.

Tilt (Inclination of Equator Relative to Orbital Plane). 29.6°.

Atmospheric Composition. Mostly hydrogen, with some helium, a small amount of methane, and trace amounts of other gases.

General Composition. Melted ices, silicates, and hydrogen and helium.

Weight of Person Who Weighs 100 pounds (45 kilograms) on Earth.* 113.8 pounds (51.6 kilograms).

Number of Known Moons. 13.

*Calculated at the altitude at which 1 bar of atmospheric pressure (the pressure of Earth's atmosphere at sea level) is exerted.

†Sidereal revolution period, or the time it takes the planet to revolve around the sun once, relative to the fixed (distant) stars.

‡Sidereal rotation period, or the time it takes the planet to rotate about its axis once, relative to the fixed (distant) stars.

a planet from the time of its discovery in 1930 until 2006, when it was reclassified as a dwarf planet. So, for some 75 years, Neptune was considered the second farthest planet from the sun.

Like the other giant planets, Neptune spins quickly, completing one rotation in about 16 hours. Its rapid rotation slightly distorts its shape, making it bulge a bit at the equator and flatten a bit at the poles. Neptune's rotational axis is tilted about 29.6 degrees relative to the plane in which it orbits. As on Earth, the axial tilt gives it seasons. As the planet travels along its orbit, first one hemisphere then the other is tipped closer to the sun. Since Neptune's orbit is almost a perfect circle, its seasons are of even length, with each season lasting about 41 Earth years.

Atmosphere

Like the other outer planets, Neptune has a massive atmosphere, or surrounding layer of gases, composed mostly of hydrogen with some helium. Methane makes up most of the rest, accounting for about 2 percent of the molecules within the atmosphere. The methane gives Neptune its bluish color. Methane strongly absorbs red light, so the light reflected off the planet's clouds lacks red and appears blue. Uranus has a similar percentage of methane, but it appears blue-green. Some other substance that has not yet been identified must give Neptune its more vivid blue color.

Temperatures vary with altitude in Neptune's atmosphere. At a point where the pressure is 1 bar, the temperature is about –326° F (–199° C). Above that point, it gets colder with altitude in a middle layer but warmer with altitude in a higher layer. The highest levels of Neptune's atmosphere are quite hot, with a constant temperature of some 890° F (480° C).

Air currents in the atmosphere probably rise at middle latitudes and descend near the equator and poles. Near the cloud tops, the winds blow east-west in zones, as on Jupiter, Saturn, and Uranus. Near Neptune's equator, the winds whip westward at some 1,570 miles (2,520 kilometers) per hour, the fastest wind speed ever found in the solar system. At higher latitudes, the winds blow eastward at slower speeds.

Voyager 2 detected considerable atmospheric turbulence during its 1989 flyby. An enormous, whirling, Earth-sized storm system called the Great Dark Spot appeared as a dark oval in photographs of the southern hemisphere. A smaller dark spot and a bright, fast-moving cloud called Scooter also appeared. Unlike Jupiter's Great Red Spot, which is at least 300 years old, Neptune's large storm systems do not seem to be long lasting. The Great Dark Spot did not show up in images made with the Hubble Space Telescope just a couple of years after the Voyager flyby. Another dark

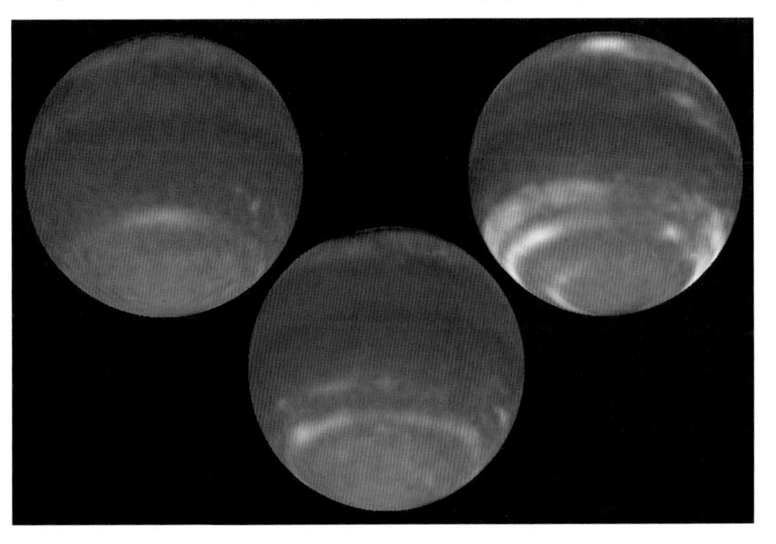

Seasonal changes in Neptune's atmosphere appear in images taken over six years during late spring in the southern hemisphere. Views made with the Hubble Space Telescope in (from left to right) 1996, 1998, and 2002 show bands of clouds over the southern hemisphere becoming wider and brighter as that part of the planet receives increasingly more sunlight. The images were taken in visible and near-infrared light.

NASA, L. Stromovsky, and P. Fry (University of Wisconsin-Madison)

Photo NASA/JPL/Caltech (NASA photo # PIA00058)

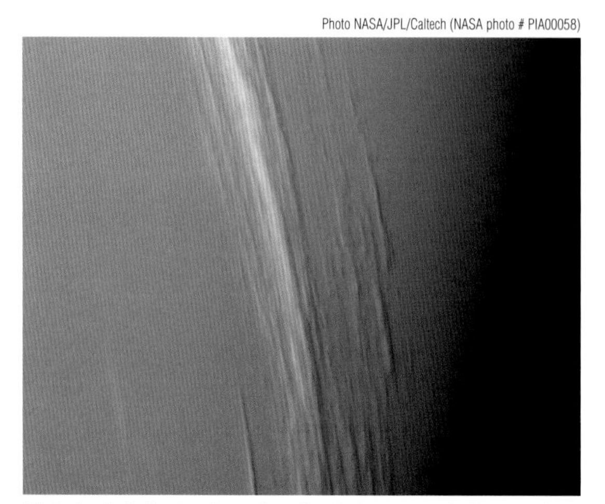

High clouds in Neptune's northern hemisphere cast shadows on the main cloud deck below in an image taken by Voyager 2. The cloud bands stretch latitudinally. They are illuminated by sunlight from the left.

spot appeared for a few years in the planet's northern hemisphere in Hubble images taken in the 1990s.

The highest clouds in Neptune's atmosphere appear in delicate bands like Earth's cirrus clouds. They are probably formed of crystals of frozen methane. These high, dispersed clouds cast shadows on the main cloud deck below, which may be formed of crystals of frozen ammonia or hydrogen sulfide. Cloud layers probably exist below the main deck, including some likely made of water ice.

Interior

The pressures and temperatures inside Neptune are very high, so the interior is probably liquid. Scientists think it is made mostly of melted ices of water, methane, and ammonia plus molten silicate rock and metal. It also contains a smaller percentage of hydrogen and helium. Neptune's composition is most similar to that of Uranus. Since Neptune is denser, it is thought to contain a greater percentage of the heavier substances—the molten rocky material and melted ices. Scientists doubt that Neptune has distinct layers like Jupiter and Saturn. Instead, the heavier substances are probably well mixed throughout the fluid interior, as in Uranus.

Like Jupiter and Saturn, Neptune radiates roughly twice as much energy as it gets from the sun, for some unknown reason. It seems to have a strong internal heat source. The planet is more than a billion miles (1.6 billion kilometers) farther from the sun than Uranus is, and it receives less than half the sunlight that Uranus does. Nevertheless, Neptune is slightly warmer than its inner neighbor.

Neptune produces its own magnetic field, which has a north pole and a south pole, like a bar magnet. However, the north magnetic pole and the north rotational pole are skewed. Neptune's magnetic field is tilted almost 47 degrees relative to its rotation axis. Only Uranus' magnetic field is tilted more. This configuration may indicate that processes in the upper layers of the interiors, and not the centers, generate the magnetic fields in both planets.

Ring System

A system of six narrow rings encircles Neptune. They are made mostly of dust-sized particles, all orbiting the planet. Arcs of brighter material punctuate part of the outermost and densest ring, called Adams. The rings are named after astronomers who discovered the planet and its important features. Inward from Adams, which lies some 39,000 miles (63,000 kilometers) from Neptune's center, are Galatea, Arago, Lassell, Le Verrier, and Galle, the innermost ring at about 26,000

Major Moons of Neptune

	Naiad	Thalassa	Despina	Galatea
average distance from center of planet	29,970 mi (48,230 km)	31,115 mi (50,075 km)	32,640 mi (52,525 km)	38,500 mi (61,955 km)
diameter or dimensions	60 x 37 x 32 mi (96 x 60 x 52 km)	67 x 62 x 32 mi (108 x 100 x 52 km)	112 x 92 x 80 mi (180 x 148 x 128 km)	127 x 114 x 89 mi (204 x 184 x 144 km)
orbital period	0.3 Earth days	0.3 Earth days	0.3 Earth days	0.4 Earth days
rotation period	likely synchronous (same as orbital period)	likely synchronous (same as orbital period)	likely synchronous (same as orbital period)	likely synchronous (same as orbital period)
discovery year, discoverer	1989, identified in Voyager 2 images	1989, identified in Voyager 2 images	1989, identified in Voyager 2 images	1989, identified in Voyager 2 images

	Larissa	Proteus	Triton	Nereid
average distance from center of planet	45,700 mi (73,550 km)	73,100 mi (117,650 km)	220,500 mi (354,800 km)	3,425,900 mi (5,513,400 km)
diameter or dimensions	134 x 127 x 104 mi (216 x 204 x 168 km)	273 x 258 x 251 mi (440 x 416 x 404 km)	1,681 mi (2,706 km)	210 mi (340 km)
orbital period	0.6 Earth days	1.1 Earth days	5.9 Earth days (retrograde)	360.1 Earth days
rotation period	likely synchronous (same as orbital period)	likely synchronous (same as orbital period)	synchronous (same as orbital period)	not synchronous (not the same as orbital period)
discovery year, discoverer	1989, identified in Voyager 2 images	1989, identified in Voyager 2 images	1846, William Lassell	1949, Gerard P. Kuiper

NASA/JPL

Three bright arcs appear in Adams, the outermost ring of Neptune, in an image of part of the planet's ring system taken by Voyager 2.

miles (42,000 kilometers) from the planet's center. The four innermost moons of Neptune orbit within the ring system. At least some of them may be shepherd moons, whose gravity keeps the rings from spreading out.

NASA/JPL

Triton, the largest moon of Neptune, appears in a composite of 14 images taken by Voyager 2. A large ice cap, presumably of frozen nitrogen, covers the southern polar region. Terrain that looks like the rind of a cantaloupe appears to the north of the ice cap.

Moons

Neptune has 13 known moons. The largest, Triton, was discovered in 1846 only about a month after Neptune was discovered. With a diameter of 1,681 miles (2,706 kilometers), it is a bit smaller than Earth's moon but a bit larger than Pluto. Triton is also similar to Pluto in density—both are about twice as dense as water—and surface composition. Frozen methane and nitrogen cover the moon's surface. At about –390° F (–235° C), it is one of the coldest known surfaces in the solar system. Triton's very tenuous atmosphere is composed mostly of nitrogen. Voyager 2 captured images of large, geyserlike plumes erupting. These active "ice volcanoes" likely spew nitrogen gas and large dust particles.

Unlike all other large moons in the solar system, Triton revolves around Neptune in a direction opposite to the planet's rotation. Also, Triton's orbit is tilted more than 157 degrees relative to Neptune's equator; most large moons are inclined less than about 5 degrees. These peculiarities suggest that Triton may have formed as an icy object elsewhere in the solar system and was later captured by Neptune's gravity. As its path was adjusted into a circular orbit, the pull of Neptune's gravity probably caused the moon's interior to melt and separate into layers. Triton later refroze.

The distant moon Nereid was discovered in 1949, also by telescope. It has the most eccentric orbit of any known moon in the solar system. The moons Naiad, Thalassa, Despina, Galatea, Larissa, and Proteus were not known until Voyager 2 visited the planet in 1989. They travel in nearly circular orbits that lie near the plane of Neptune's orbit. The other five moons were discovered by telescope in 2002–03. They are tiny outer moons with eccentric orbits that lie outside the plane of Neptune's orbit.

Spacecraft Exploration

The unmanned United States Voyager 2 probe is the only spacecraft to have visited far-off Neptune and its moons. After its launch in 1977, the craft visited Jupiter and Saturn and was then reprogrammed so that it could visit Uranus and finally Neptune. Along the way, Voyager used the gravitational fields of each of the planets to boost its speed and divert it toward the next planet on its course. In 1989 the craft flew past Neptune, observing the planet and its moons for several months. Its closest approach took it only about 3,100 miles (5,000 kilometers) above the planet's north pole. The data and the some 10,000 photographs it obtained substantially increased knowledge about Neptune and Triton, confirmed that the planet has rings, and revealed for the first time six of its moons.

Exploring Neptune: Spacecraft Missions

Spacecraft	Type of Mission	Launch Date	Country or Agency	Highlights
Voyager 2	flyby	Aug. 20, 1977	U.S.	First and only spacecraft to visit Uranus and Neptune. Took advantage of rare planetary alignment to fly by all four outer planets. Flew past Neptune on Aug. 24–25, 1989, uncovering the rings, six moons, Neptune's magnetic field, and surprisingly high-speed winds and active storms in its atmosphere. After the Neptune flyby, the craft was hurled toward interstellar space.

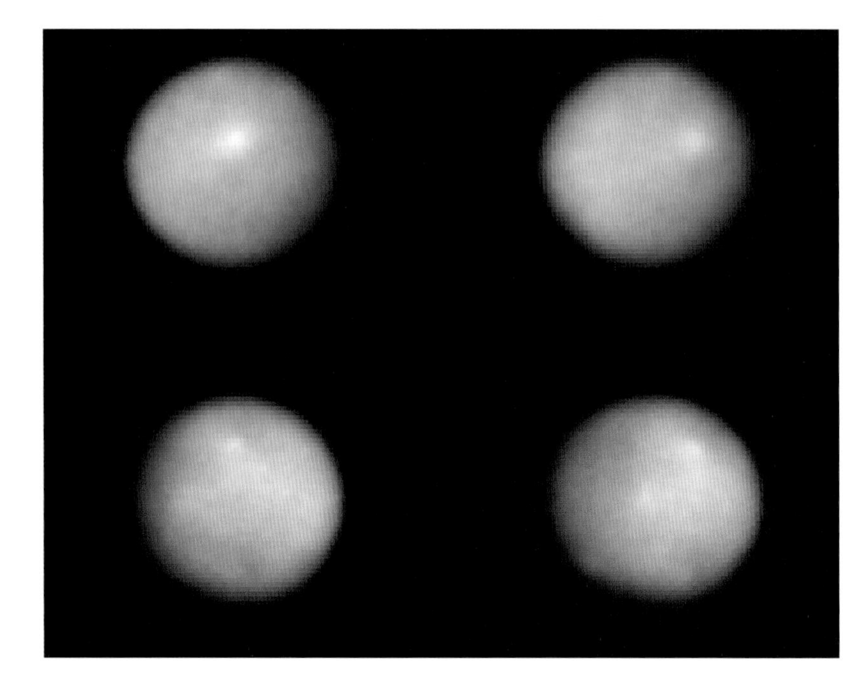

Four false-color images show the rotation of Ceres, a dwarf planet and the largest asteroid. Like the eight planets, Ceres is massive enough for its gravity to have molded it into a round shape. Each image is a composite of many exposures taken by the Hubble Space Telescope in visible and ultraviolet light. Ceres' small size and distance from Earth make it difficult to photograph from Earth's vicinity.

ESA/STScI/NASA

DWARF PLANET

The objects called dwarf planets are similar to the solar system's eight planets but are smaller. Like planets, they are large, roundish objects that orbit the sun but that are not moons. The first three objects classified as dwarf planets, in 2006, were Pluto, Eris, and Ceres. Makemake (pronounced "mah-kay mah-kay") and Haumea were named dwarf planets in 2008.

The category dwarf planet was created as a result of intense debate as to whether Pluto should be called a planet. Pluto had been considered the solar system's ninth planet at the time of its discovery in 1930. However, in 2006 the International Astronomical Union (IAU), the organization responsible for approving the names of astronomical objects, defined "planet" so that only eight bodies in the solar system qualified. At the same time, it established a new, distinct class of objects called dwarf planets. According to the IAU, both planets and dwarf planets must orbit the sun and be massive enough for their gravity to have pulled them into spherical or nearly spherical shapes. Whether such an object is classified as a planet or dwarf planet depends on whether it has "cleared the neighborhood around its orbit." An object with a mass great enough for its strong gravity to have swept up or deflected away most of the smaller nearby bodies is considered a planet. An object that has failed to do so, and thus failed to grow larger, is a dwarf planet.

Eris, Pluto, Makemake, and Haumea are large members of the Kuiper belt, a distant region containing countless small, icy bodies orbiting the sun. Ceres, the largest asteroid, orbits the sun from within the main asteroid belt. So these five bodies do not qualify as planets under the IAU's definition because they have not cleared away many chunks of icy and rocky debris from their orbital vicinities.

For practical purposes, objects classified as dwarf planets are smaller than the planet Mercury, which has a diameter of about 3,032 miles (4,879 kilometers). Eris is thought to have a diameter of roughly 1,550 miles (2,500 kilometers). Pluto is slightly smaller, with a diameter of about 1,456 miles (2,344 kilometers). With a diameter of perhaps about 1,000 miles (1,600 kilometers), Makemake is thought to be some two thirds the size of Pluto. Ceres is the smallest dwarf planet, with a diameter of about 584 miles (940 kilometers). Haumea is an unusual object. Although it is substantially rounded, it is also quite elongated. It rotates about its axis so quickly—completing one rotation in just under four hours—that it is pulled into a shape somewhat like an American football. Its longest dimension is thought to be roughly the size of the diameter of Makemake or Pluto.

In 2008 the IAU decided on a name for a new subcategory of dwarf planets—plutoids. A plutoid is a dwarf planet whose orbit takes it farther from the sun than Neptune, on average. To be named as a plutoid, a dwarf planet must also meet a brightness requirement (an absolute magnitude brighter than +1). Pluto, Eris, Makemake, and Haumea are considered to be both dwarf planets and plutoids. Ceres, which orbits much closer to the sun, is a dwarf planet but not a plutoid. (*See also* Planet; Pluto; Eris; Asteroid.)

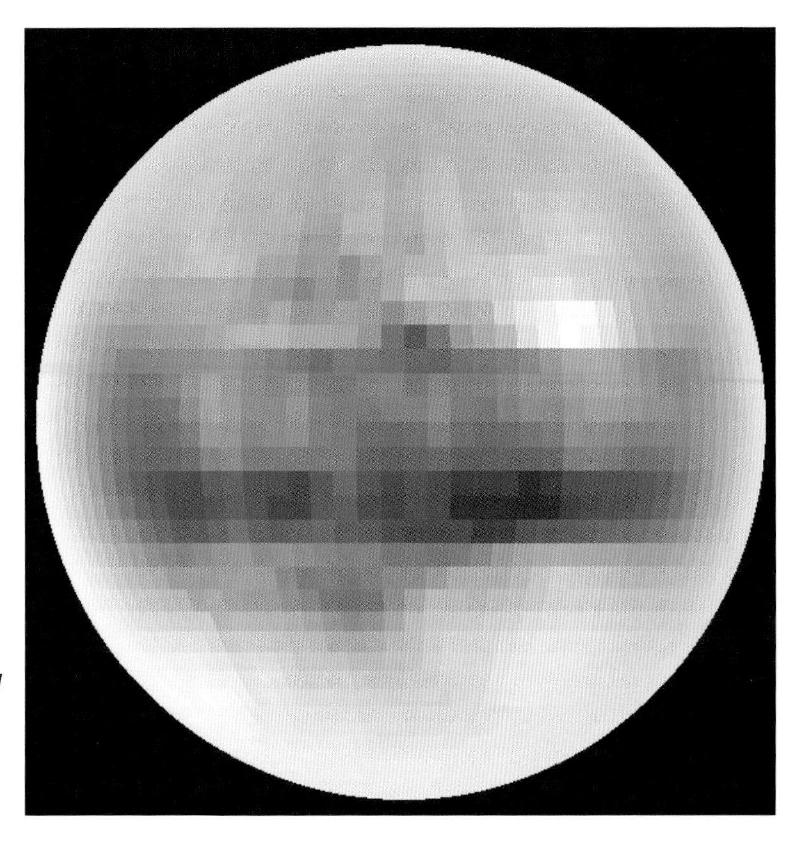

Because Pluto is so far from Earth, it is difficult to capture images of it. A false-color map of the distant world was created from telescopic data collected between 1985 and 1990, during a period of multiple mutual eclipses between Pluto and its largest moon, Charon.

Eliot Young, Southwest Research Institute; NASA's Planetary Astronomy Program

PLUTO

The distant rocky and icy body named Pluto is a dwarf planet. For 76 years, however, from its discovery in 1930 until 2006, it was considered the ninth and outermost planet of the solar system. Pluto is on average about 39.5 times farther from the sun than Earth is. As a result, very little sunlight reaches Pluto, so it must be a dark, frigid world. Fittingly, it was named after the god of the underworld in ancient Roman mythology. It has three moons, two of which are tiny. Its largest moon, Charon, is so large with respect to Pluto that the two are often considered a double-body system.

Scientists know less about Pluto than they do about the planets. Because Pluto is so far from Earth, its features are difficult to observe with even the most powerful telescopes. Key observations have been made using instruments that orbit Earth from above its atmosphere, including the Hubble Space Telescope. Earth-based telescopes outfitted with equipment to reduce the blurring effects of the atmosphere have also been effective. The first spacecraft mission to Pluto, New Horizons, is expected to dramatically increase knowledge about the distant world. The United States National Aeronautics and Space Administration (NASA) launched the unmanned probe in 2006 on a nine-year trip to Pluto and Charon.

Eclipses between Pluto and Charon have helped scientists learn more about the two bodies. From 1985 to 1990 the two eclipsed each other about once every 6.4 days. As seen from Earth, Charon would alternately cross directly in front of Pluto, partly blocking it from view, or be hidden behind it. By carefully observing these events, scientists were able to calculate the size and mass of Pluto and Charon much more accurately than before and to make a crude map of Pluto's surface.

Discovery and Classification

Pluto is too far from Earth to be visible with the unaided eye. It was discovered in the early 20th century when astronomers began searching the skies for a new planet beyond the eight planets then known. They had noted what seemed to be irregularities in the orbits of Uranus and Neptune, and they thought that the gravity of a planet beyond Neptune might be the cause. In 1929 the Lowell Observatory hired Clyde Tombaugh, a 23-year-old amateur astronomer, to continue the search. He discovered Pluto in 1930, and it was considered a planet. Pluto turned out to be much too small, however, to account for the orbital disturbances of Uranus and Neptune (which more-accurate data later showed not to exist anyway). The discovery of Pluto was thus serendipitous—the result of a lucky accident.

Pluto proved to be unlike any other planet in orbit, size, density, and composition. In the late 20th century astronomers began to discover many more small icy objects that orbit the sun from beyond Neptune, in a region called the Kuiper belt. Pluto seemed to have more in common with those objects than with the planets. Most of the Kuiper belt objects discovered were smaller than Pluto. A few were close to it in size, and one—later named Eris—was slightly larger. Many astronomers proposed that either Pluto should not be called a planet or that some other large celestial objects should also be classified as planets.

In 2006 the International Astronomical Union (IAU), the group that approves the names of astronomical objects, defined "planet" so that Pluto was excluded. It fails to meet one of the IAU's new criteria: having cleared other objects from the vicinity of its orbit. Instead, the IAU designated Pluto as the prototype of a new category of celestial objects, dwarf planets. In 2008 the IAU named a new subcategory of dwarf planet—plutoids—after Pluto. Pluto and other plutoids are bright dwarf planets that are on average farther from the sun than Neptune is. Pluto is also considered a Kuiper belt object. (*See also* Planet; Dwarf Planet.)

Basic Astronomical Data

Pluto is one of the largest members of the Kuiper belt, but it is considerably smaller than the solar system's planets. In size, density, and composition, it most resembles Triton, the large icy moon of Neptune. The diameter at Pluto's equator is about 1,456 miles (2,344 kilometers), which is less than half that of the smallest planet, Mercury. Several moons, including Earth's moon and Triton, are larger. Triton has about twice the mass of Pluto, and Mercury has more than 27 times the mass. Like Triton, Pluto is only about twice as dense as water. Its low density suggests that it is made of a high percentage of ice as well as rock. By comparison, Mercury and Earth are both more than five times as dense as water.

Like the eight planets, Pluto revolves around the sun in an elliptical, or oval-shaped, orbit. However, Pluto's orbit is more tilted and more eccentric, or elongated, than the orbits of the planets. While the planets orbit in about the same plane as Earth does, Pluto's orbit is inclined about 17° from that plane. Mercury has the most tilted orbit of the planets, with an inclination of about 7°.

Pluto's orbit is quite elongated. As a result, as it travels along its path, its distance from the sun varies considerably, from roughly 2.8 billion to 4.6 billion miles (4.5 billion to 7.3 billion kilometers). For a small part of its orbit, Pluto is actually closer to the sun than Neptune is. The last time this happened was in 1979–99. Pluto and Neptune will never collide, however. In the time it takes Neptune to revolve around the sun three times, Pluto revolves only twice, in such a way that the bodies never pass each other closely. Since Pluto is so distant, it takes nearly 248 Earth years to complete just one trip around the sun. In other words, a year on Pluto is about 248 times longer than one on Earth.

Scientists studying Pluto through telescopes have observed that its brightness level varies regularly in a period of about 6.4 days. This variation indicates that some areas of the surface reflect more light than others and that Pluto completes one rotation on its axis in about 6.4 days. In other words, one day on Pluto lasts about 6.4 Earth days.

Pluto's rotational axis is tipped about 120° relative to its orbital plane, so that, like Uranus, it lies nearly on its side. Both those bodies spin in retrograde motion, or the direction opposite that of most of the planets. An observer on Pluto would see the sun rise in the west and set in the east.

Pluto's large, eccentric orbit and the great tilt of its axis must give it long, uneven, and extreme seasons. The amount of sunlight reaching the dwarf planet varies greatly during its year, since its distance from the sun varies so much. Since its axis is so tilted, during parts of its orbit it points its north pole almost directly at the sun. It is then dark in the southern hemisphere day and night. When Pluto is on the opposite side of its orbit, the situation is reversed, with the northern hemisphere in constant darkness.

Atmosphere, Surface, and Interior

In 1988 Pluto passed directly in front of a star as seen from Earth, an event called an occultation. The way

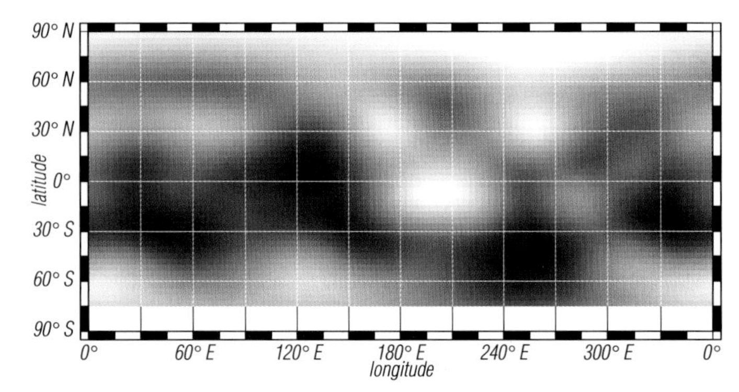

Bright and dark regions on Pluto's surface appear in a map based on images taken with the Hubble Space Telescope. The north polar region generally has bright areas, while the equatorial region, particularly to the south, has more dark patches. The map is a Mercator projection; the blue tint was added in reproduction.

Source image: AURA/STScI/NASA/JPL

the star's light gradually dimmed before being temporarily blocked out completely to Earth-based observers showed scientists that Pluto has an atmosphere, or surrounding layer of gases. It is very tenuous and extends far above the dwarf planet.

The atmosphere is composed of nitrogen, with smaller amounts of methane and carbon monoxide. The density of the gases is always low, but it must vary as Pluto's distance from the sun varies. At its closest to the sun, Pluto is warmer. Some of the frozen gases on its surface must then vaporize, becoming gases in the air. As Pluto moves away from the sun and gets colder, gases in the air must freeze onto the surface again. Scientists estimated that in 2000 the surface pressure was only a few microbars to several tens of microbars, with one microbar being about a millionth of the surface pressure on Earth. Pluto was then near its closest to the sun, so the atmosphere was near its densest. The atmosphere may not be detectable at all when Pluto is farthest from the sun.

Pluto's surface has some bright regions and some dark regions. Overall, it reflects about 55 percent of the light that reaches it. In comparison, Earth's moon reflects only about 10 percent of the light it receives, while icy Triton reflects about 80 percent, as ice is highly reflective. Pluto's fairly high reflectivity suggests that its surface consists partly of ices and partly of something else. The brighter regions seem to be mostly frozen nitrogen, with some frozen methane, water ice, and frozen carbon monoxide. The area around the south pole is especially bright. Little is known about the darker regions of the surface, which are somewhat reddish. It is thought that they contain a mixture of organic compounds.

Because it is so far from the sun, Pluto receives only about 1/1,600 of the amount of sunlight that Earth does, on average. Its surface is very cold. Different types of observations have suggested that the surface temperature may be about –355° F to –397° F (–215° C to –238° C). The temperature probably varies seasonally, and the brighter areas must be generally colder than the darker ones.

Pluto is thought to be made of more than half rock with the rest ice, probably water ice. Scientists think its interior may have separated into layers, with a rocky core surrounded by a mantle of water ice, but more information is needed.

Moons

In the years after Pluto's discovery, attempts to detect moons were unsuccessful because Pluto is so remote. Moreover, Charon, its only major moon, lies unusually close to Pluto. It is only about 12,200 miles (19,640 kilometers) from Pluto's center, or nearly 20 times closer than Earth's moon is to Earth. As a result, Charon was obscured by the glare of Pluto's light. In 1978 James W. Christy and Robert S. Harrington of the U.S. Naval Observatory discovered Charon while examining photographs of Pluto. The moon was named after the boatman in ancient Greek mythology who ferried souls across the river Styx to Hades (the Greek counterpart of Pluto) for judgment. (Codiscoverer Christy also had his wife, Charlene, in mind in naming the moon and wanted its initial sound to be pronounced "sh" rather than "k.")

HST Pluto Companion Search/ESA/NASA

Pluto and its three known moons appear in an image taken by the Hubble Space Telescope.

For a moon, Charon is very large relative to Pluto. Its diameter is about 750 miles (1,210 kilometers), which is about half that of Pluto. Its density is about 1.7 times that of water, and it reflects about 35 percent of the light that hits it. It is thought to be a bit more than half rock, with the rest ice. Unlike Pluto, Charon probably has water ice covering much of its surface. Scientists were able to refine their estimates of Charon's size and density after they observed it occulting a star in 2005. The occultation also indicated that the moon either does not have an atmosphere or has an extremely tenuous one (much less massive even than Pluto's).

Charon completes one revolution around Pluto and one rotation on its axis in about 6.4 days. It thus rotates synchronously, so that the same hemisphere is always facing Pluto. Because Pluto's rotational period also is about 6.4 days, Pluto in turn always faces its same hemisphere toward Charon.

Scientists think that Pluto formed when the solar system condensed out of a gaseous cloud some 4.6 billion years ago. A collision between Pluto and a large body is thought to have knocked debris into a ring around Pluto, and the debris clumped together into an object that ultimately developed into Charon. Earth and its large moon probably formed through a similar collision.

A team of astronomers discovered two much smaller moons in 2005 in images taken with the Hubble Space Telescope. The moons, named Nix and Hydra, may be roughly 35 miles (55 kilometers) in diameter. They may have formed from the same collision that created Charon.

The distant dwarf planet Eris orbits the sun from more than 6 billion miles (10 billion kilometers) away, on average. Its remoteness makes it very difficult to observe from the vicinity of Earth, even with the most powerful telescopes. An image taken at infrared wavelengths at the W.M. Keck Observatory (left) shows Eris and its moon, Dysnomia—the point of light to the right of Eris. An image taken by the Hubble space Telescope (above) shows Eris in visible light.

(Left) W.M. Keck Observatory; (right) ESA/NASA

ERIS

The object named Eris orbits the sun from well beyond the orbits of Neptune and Pluto. It is the largest-known member of the Kuiper belt, a very distant, doughnut-shaped zone of countless icy objects orbiting the sun. Eris is also classified as a dwarf planet, being massive enough for its gravity to have pulled it into a spherical shape but not as massive as the eight planets of the solar system. It is also a plutoid—a bright dwarf planet that is farther from the Sun than Neptune is, on average.

Eris' diameter of roughly 1,550 miles (2,500 kilometers) makes it slightly larger than its fellow dwarf planet and Kuiper belt object Pluto. Eris is even more remote than Pluto, being on average nearly 68 times farther from the sun than is Earth, compared with 39.5 times for Pluto. Scientists calculate that it takes Eris some 560 Earth years to complete just one revolution around the sun. Its orbit is highly eccentric, or elongated, and very tilted. Its orbit is so tilted that some scientists call it a member of the Kuiper belt's scattered disk rather than of the Kuiper belt itself. The surface of Eris may be covered with white methane ice. Eris has one known moon, Dysnomia. The moon is about an eighth as big as Eris and takes about two weeks to circle the dwarf planet.

Eris was discovered in 2005 in images taken two years earlier at Palomar Observatory in the U.S. state of California by Michael E. Brown, Chad Trujillo, and David Rabinowitz. The Kuiper belt object was provisionally designated 2003 UB_{313} and was nicknamed "Xena" (after a character in a television series) and "the 10th planet" before receiving its official name, Eris, in 2006. Eris was named after the goddess of discord and strife in ancient Greek mythology. The name is fitting, since the discovery of the celestial object led to a great controversy in the world of planetary science. Since Eris is larger than Pluto, some scientists thought that it, too, should be considered one of the solar system's major planets. Others preferred a classification scheme in which both Pluto and Eris are excluded; such a scheme was officially adopted in 2006 (*see* Planet).

Eris' moon, Dysnomia, was discovered in 2005 in infrared images taken at the W.M. Keck Observatory in the U.S. state of Hawaii. The moon was named for the daughter of Eris in Greek mythology, who was associated with lawlessness.

NASA/JPL/Caltech

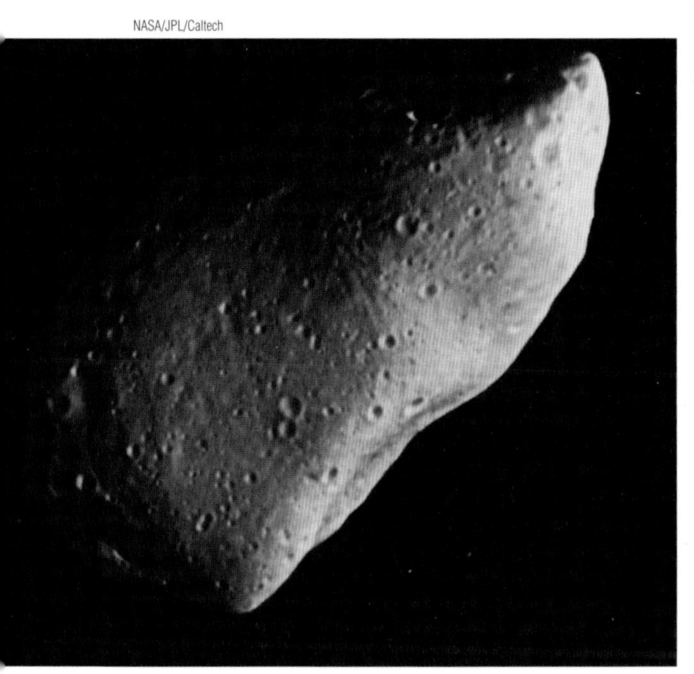

Numerous craters mark the surface of Gaspra, an asteroid of the main belt, which appears in a composite of two images taken by the Galileo spacecraft. The colors were enhanced by computer to highlight subtle variations in surface properties.

ASTEROID

The many small bodies called asteroids are chunks of rock and metal that orbit the sun. Most are found in the main asteroid belt, a doughnut-shaped zone between the orbits of Mars and Jupiter. Astronomers think that when the solar system was forming, the pull of gravity from the object that became Jupiter prevented the asteroids from clumping together to form a planet. The discovery of asteroids dates to 1801, when astronomer Giuseppi Piazzi found Ceres.

Asteroids are also called minor planets because they are smaller than the major planets of the solar system. Ceres is the largest asteroid by far, with a diameter of about 584 miles (940 kilometers). It is massive enough to also be considered a dwarf planet. Only about 30 asteroids are greater than 125 miles (200 kilometers) in diameter. Most are much smaller. There are probably millions of boulder-sized asteroids in the solar system. These small objects likely result from collisions of larger asteroids. The largest asteroids have enough mass for their gravity to have pulled them into nearly spherical shapes, but smaller asteroids can be elongated or irregularly shaped.

Astronomers classify asteroids into more than a dozen categories by the fraction of sunlight they reflect and by the spectrum (or color) of the reflected light, which provide clues to the surface composition. There are three broad types: C, S, and M. Most large- and medium-sized asteroids are C-type, or carbonaceous, asteroids. They are very dark, containing a considerable amount of carbon and complex organic compounds. C-type asteroids predominate in the outer part of the main belt. S-type, or silicaceous, asteroids are relatively bright stony bodies. Their surfaces contain silicate rock with some iron. S-type asteroids are found mainly in the inner regions of the main belt. M-type, or metallic, asteroids also are fairly bright. Their surfaces probably contain much iron. They occur in the middle of the main belt.

Most asteroids occur in groups within the main belt. However, many asteroids, called Trojans, orbit in two clusters at Jupiter's distance from the sun, with one group about 60 degrees ahead of Jupiter and the other about 60 degrees behind. The Trojans are very dark, like C-type asteroids, but many have a distinctly dark-reddish hue. They are rich in organics, may have ice deep within, and are thought to resemble comet nuclei.

Other asteroids approach the sun more closely than do those in the main belt. Those that pass close to Earth's orbit—within roughly 28 million miles (45 million kilometers)—are called near-Earth asteroids. The orbits of some of these asteroids cross Earth's orbit, so one could collide with Earth in the future. Several programs are dedicated to identifying and studying these potentially hazardous asteroids.

Small asteroids and asteroid fragments regularly strike Earth's surface in the form of meteorites. Much less often, large asteroids crash into Earth, forming huge craters. Past large impacts may have caused earthquakes, giant sea waves, and even global dust clouds that blocked sunlight for long periods. One theory put forth to explain the mass extinction of the dinosaurs and other species some 65 million years ago is that a large asteroid slammed into Earth, causing a massive disturbance in the global climate. A crater at Chicxulub, in southeastern Mexico, is thought to have been created by that asteroid.

About 20 percent of near-Earth asteroids have been found to be double bodies or to have satellites (moons) orbiting around them. Smaller percentages of main-belt and Trojan asteroids have satellites.

The first asteroidal satellite was discovered by the Galileo spacecraft, which provided the first close-up views of asteroids. The craft flew by two S-type main-belt asteroids—Gaspra in 1991 and Ida in 1993. Ida was found to have a moonlet, Dactyl, which measures about a mile (1.5 kilometers) across. Both Ida and Dactyl appear to have the composition of ordinary chondrite meteorites, the most common kind in meteorite collections. The Near Earth Asteroid Rendezvous (NEAR) Shoemaker probe was the first to fly past a C-type asteroid, Mathilde, in 1997. It later became the first craft to conduct a long-term study of an asteroid at close range, orbiting the S-type near-Earth asteroid Eros for about a year in 2000–01. The first craft to land on a small body, it made a controlled descent onto Eros' surface in 2001. One important discovery was that Eros never underwent extensive melting and separation into layers, so it may be a pristine sample of primordial solar system material.

Comet Hale-Bopp blazes across the sky in an image taken from Earth in 1997. The comet has a blue tail of plasma (hot, ionized gases) and a white tail of dust.

COMET

When near the sun, the small bodies called comets develop a hazy cloud of gases and dust. They also often develop long, glowing tails. However, a comet exists as only a small core of ice and dust for most or even its entire orbit around the sun. Comets can only be easily seen from Earth when they approach the sun closely. Even then, most are visible only with a telescope. Among the exceptionally bright "naked eye" comets seen from Earth after 1900 were the Great Comet of 1910, Halley's, Skjellerup-Maristany, Seki-Lines, Ikeya-Seki, Arend-Roland, Bennett, West, Hyakutake, Hale-Bopp, McNaught, and Holmes. When comets are far from the sun, they appear in large telescopes as a point of light, like a star.

Most comets originate in the very distant, outer regions of the solar system. They are thought to be the remnants of the building blocks that produced the planets Uranus and Neptune some 4.6 billion years ago. Comets remain essentially unchanged when they are away from the sun in the deep cold of space, which for many comets can be for eons. For this reason, astronomers think that comets may contain some of the oldest and best-preserved material in the solar system.

Many people look forward to sighting a comet, but for many centuries comets were believed to have an evil influence on human affairs. They were thought to foretell plagues, wars, and death. It was not until the 17th century that they began to be properly understood. Astronomer Edmond Halley studied the written accounts of 24 comets that had been seen from 1337 to 1698 and calculated their orbits. He found that the comets of 1531, 1607, and 1682 moved in almost the same paths, and he concluded that they were all the same comet, which would return in about 1758.

His forecast was correct, for the comet did appear in that year, though Halley himself did not see it; he died in 1742. For the first time scientists realized that comets can be regular visitors, and the comet was named after Halley. It takes the comet about 76 years to complete each trip around the sun—which is fairly short for a comet. Last appearing in 1986, Halley's comet will not be seen again until 2061.

Structure and Composition

The only permanent part of a comet is its solid nucleus, or core. It is typically very dark, irregularly shaped, and several miles in diameter. The nucleus is often described as a "dirty snowball" because it consists of ice mixed with large amounts of fine, sooty dust particles. Some comets may have more dust than ice. The ice in a comet is mainly frozen water, with smaller amounts of frozen carbon monoxide, carbon dioxide, methane, ammonia, and other frozen gases. The dust contains rocky material and organic compounds. Comet nuclei are fragile and have been observed breaking up into fragments.

As a comet nears the sun, the ice in the nucleus begins to sublimate—that is, to pass directly from a solid to a gas. The gas carries with it some of the loosely bound dust particles. The gases spread out around the nucleus, forming a huge, dusty atmosphere called the coma. The nucleus and coma together make up the head of the comet. The diffuse, gaseous coma is what makes the head of a comet appear hazy. The coma is enormous, typically reaching about 60,000 miles (100,000 kilometers) or more in diameter. Sunlight causes the atoms in the coma to glow. If the supply of gases from the nucleus changes, a comet can brighten or fade unexpectedly, so astronomers cannot predict how bright a comet will become.

As a comet approaches the sun, radiation from the sun usually blows dust from the comet into a dust tail. This tail is typically wide, slightly curved, and yellowish. The solar wind, a stream of highly energetic charged particles from the sun, often sweeps hot gases away in a slightly different direction, producing another tail. It is usually fairly narrow, straight, and bluish. This tail is formed of plasma, or gases heated so much that that they are electrically charged, with the electrons stripped away from the atomic nuclei. Comet tails may extend roughly 60 million miles (100 million kilometers) or more, but they contain only a small amount of matter. They point generally away from the sun because of the force exerted by radiation and solar wind on the cometary material. When comets travel away from the sun, therefore, their tail or tails are in front of them.

Each time a comet passes close to the sun, it loses some of its matter. Eventually, the comet may disintegrate, ending up as only a swarm of particles.

Comet Tempel 1's nucleus appears in a composite of several images taken by the Deep Impact spacecraft. The nucleus measures about 9 miles (14 kilometers) at its widest point. It has a powdery surface and a variety of terrain, including smooth and rough areas and what appear to be impact craters.

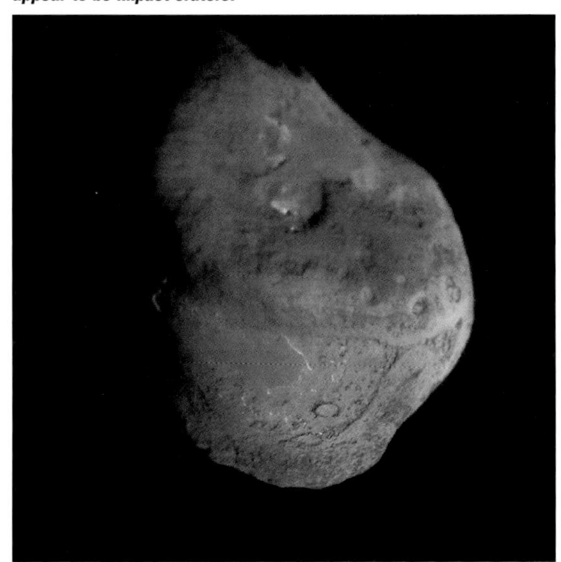

NASA/JPL/UMD

Alternately, all the ices may eventually vaporize away from near its surface, leaving a dormant, or dead, comet, which resembles an asteroid.

Orbits and Sources

Comets orbit the sun in elliptical, or oval-shaped, orbits that tend to be highly eccentric, or elongated. A comet's distance from the sun usually varies considerably along its orbit. Based on their orbits, comets can be divided into two main types: short-period comets and long-period comets.

Short-period comets take less than 200 years to complete one orbit around the sun; most of them take less than 20 years and are called Jupiter-family comets. On the other hand, long-period comets take between 200 and a million years to orbit the sun. More-distant comets—those that are more than 10,000 times farther from the sun than Earth is—have probably never been inside the planetary system before. When one is dislodged from such a great distance, perhaps as far as halfway to the nearest stars, it arrives in the inner solar system as a new comet. The orbits of long-period and new comets are often extremely elongated and greatly inclined relative to the plane in which the planets orbit. Moreover, about half of them orbit the sun in the direction opposite that of the planets. Short-period comets usually have more circular orbits that lie within the plane of the planets' orbits. Most of them orbit in the same direction as the planets.

These two different types of comets apparently come from two different sources—the distant regions called the Kuiper belt and the Oort cloud. Each of these regions is a vast reservoir of comet nuclei, consisting of countless icy small bodies orbiting the sun. Most short-period comets are thought to originate in the Kuiper belt, a doughnut-shaped region beyond Neptune. The more-distant Oort cloud is the source for most long-period comets. It is a spherical cloud of comet nuclei orbiting the sun in all directions. Sometimes the gravity of a larger body may alter the orbit of a comet nucleus in these regions, sending it on a path that takes it closer to the sun. The object then becomes a comet.

Exploration

Many spacecraft missions to comets have been highly successful. In the 1980s several probes, including the European Space Agency's Giotto, flew by Halley's comet. Giotto was the first mission to return close-up images of a comet's nucleus. The United States Stardust probe visited Comet Wild 2 in 2004, photographing its nucleus and returning particles from its coma to Earth. In 2005 the United States Deep Impact craft intentionally crashed a projectile into the nucleus of Comet Tempel 1 in order to study the ejected debris and the resulting crater, which provided clues to the comet's composition. Another opportunity to study a comet came in the 1990s, when Comet Shoemaker-Levy 9 passed near Jupiter and broke into many fragments, which crashed spectacularly into the planet. The collisions temporarily left dark spots the size of Earth in Jupiter's atmosphere.

New stars are forming from the hot gas and dust of the Orion nebula, a major "stellar nursery" only some 1,500 light-years from Earth. Our sun probably formed in a similar environment. More than 500 separate images were combined to create this mosaic.

NASA, ESA, M. Robberto (Space Telescope Science Institute/ESA) and the Hubble Space Telescope Orion Treasury Project Team

ASTRONOMY

Since the beginnings of humankind people have gazed at the heavens. Before the dawn of history someone noticed that certain celestial bodies moved in orderly and predictable paths, and astronomy—an ancient science—was born. Yet some of science's newest discoveries have been made in this same field, which includes the study of all matter outside Earth's atmosphere. From simple observations of the motions of the sun and the stars as they pass across the sky, to advanced theories of the exotic states of matter in collapsed stars, astronomy has spanned the ages.

For centuries astronomers concentrated on learning about the motions of heavenly bodies. They saw the sun rise in the east and set in the west. In the night sky they saw tiny points of light. Most of these lights—the stars—seemed to stay in the same place in relation to one another, as if they were all fastened to a huge black globe surrounding Earth. Other lights, however, seemed to travel, going from group to group of stationary stars. They named these moving points planets, which means "wanderers" in Greek.

Ancient astronomers thought that the positions of celestial bodies revealed what was going to happen on Earth—wars, births, deaths, and good fortune or bad. This system of belief is called astrology. Most scientists no longer believe in astrology, but they have found that some ancient astrologers were good at observing the motions and positions of stars and planets.

This article was contributed in part by Dr. Gerard P. Kuiper, former director of the Lunar and Planetary Laboratory, University of Arizona, and Dr. Thomas L. Swihart, Professor of Astronomy, University of Arizona, and reviewed and updated by Thomas J. Ehrensperger, physics, astronomy, and meteorology upper school instructor.

THE VISIBLE SKY

When people today look at the sky without a telescope or other modern instrument, they see basically the same things the ancient astronomers saw. During the day one can see the sun and sometimes a faint moon. On a clear night one can see stars and usually the moon. Sometimes a star may seem to be in different positions from night to night: it is really a planet, one of the "wanderers" of the ancients. The planets all circle the sun, just as Earth does. They are visible from Earth because sunlight bounces off them. The stars are much farther away. Most stars are like the sun—large, hot, and bright. They shine from their own energy.

A broad strip of dim light is also visible across the night sky. It is a clustering of faint stars known as the Milky Way. The Milky Way is part of the Milky Way galaxy—an enormous cluster of stars, of which the sun is only one member out of more than 100 billion stars. Other galaxies exist far beyond the Milky Way.

Earth in Space

The apparent westward motion of the sun, the moon, and the stars is not real. They seem to move around Earth, but this apparent motion is actually caused by Earth's movement. Earth rotates eastward, completing one rotation each day. This may be hard to believe at first because when one thinks of motion, one usually also thinks of the vibrations of moving cars or trains. But Earth moves freely in space, without rubbing

NASA

Earth floats freely in space. In this image, taken at the moon by Apollo 8 astronauts, the Americas are hidden by clouds. Asia and part of Africa are on Earth's night side.

PREVIEW

The article Astronomy is divided into the following sections:

against anything, so it does not vibrate. It is this gentle rotation, uninhibited by significant friction, that makes the sun, the moon, and the stars appear to be rising and setting.

Earth is accompanied by the moon, which moves around the planet at a distance of about 30 Earth diameters. At the same time, Earth moves around the sun. Every year Earth completes one revolution around the sun. This motion, along with the tilt of Earth's rotation axis (relative to the axis of its revolution around the sun), accounts for the changes in the seasons. When the northern half of Earth is tipped toward the sun, the Northern Hemisphere experiences summer and the Southern Hemisphere, which is tipped away from the sun, experiences winter. When Earth has moved to the other side of the sun, six months later, the seasons are reversed because the Southern Hemisphere is then tipped toward the sun and the Northern Hemisphere is tipped away from the sun.

The moon does not always look the same from Earth. Sometimes it looks round, sometimes like a thin, curved sliver. These apparent changes are called the phases of the moon. They occur because the moon shines only when the sun's light bounces off its surface. This means that only the side of the moon that faces the sun is bright. When the moon is between Earth and the sun, the light side of the moon faces away from Earth. This is called the new moon, and it is not visible from Earth. When the moon is on the other side of Earth from the sun, its entire light side faces Earth. This is called the full moon. Halfway between the new and full moons, in locations on either side of Earth, are the first quarter and the last quarter (which look like half disks as viewed from Earth).

Eclipses

In ancient times people often were terrified when the sun or the moon seemed to disappear completely when normally it would be visible. They did not understand what caused these eclipses. Eventually, astronomers reasoned that lunar eclipses (when a previously full moon at least partly disappears from the night sky) are the result of Earth passing between the moon and the sun. Earth thus casts a shadow on the moon. Similarly, solar eclipses (when the sun partly or totally disappears from the daytime sky) occur when the moon passes between Earth and the sun. The moon thus blocks the sun's light temporarily.

Eclipses occur irregularly because the plane of the moon's orbit around Earth is slightly different from the plane of Earth's orbit around the sun. The two planes intersect at an angle of about 5 degrees. This

A series of photographs have been combined to show the stages in a total solar eclipse, which occurred about 45 minutes after sunrise. The separate images were taken at about 5-minute intervals. During a solar eclipse the moon passes directly between Earth and the sun. The disk of the moon can be seen covering more and more of the rising sun until, at center, it blocks the solar disk entirely. The white glow around the black disk of the moon is the sun's corona, or outer atmosphere, which normally is obscured by the much brighter light from the sun's surface. After totality, more and more of the sun is again revealed.

Larry Landolfi/Photo Researchers, Inc.

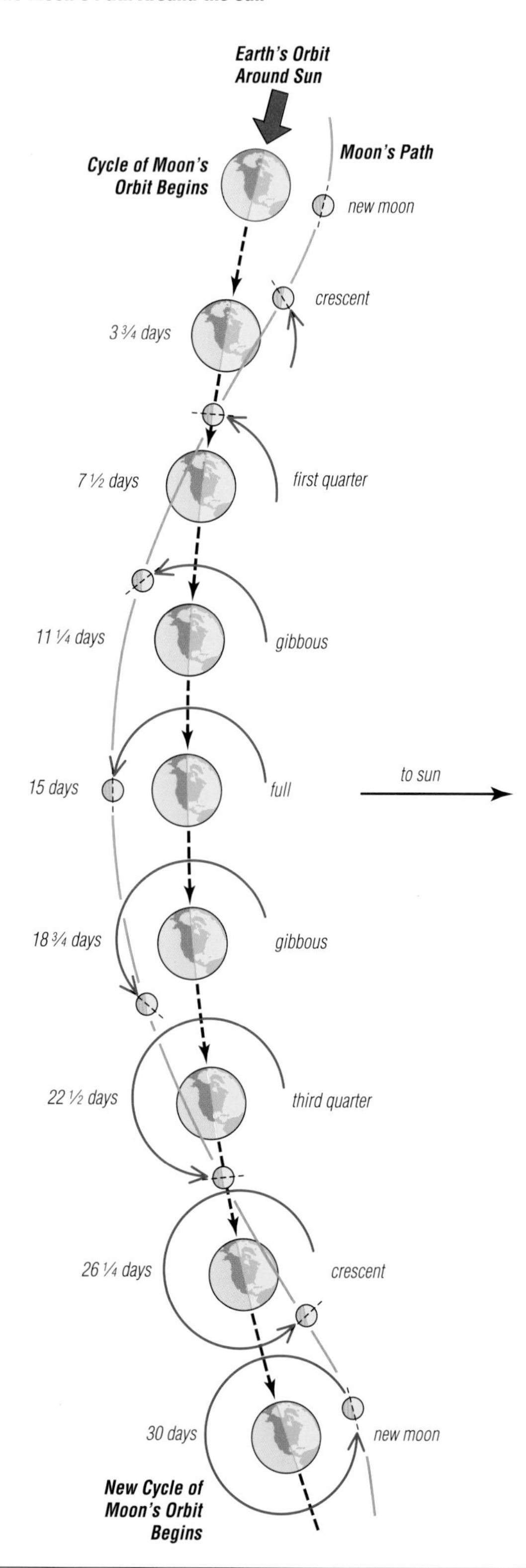

means that the moon is usually slightly above or below the line between Earth and the sun, so neither Earth nor the moon throws a shadow on the other. Eclipses can occur only when the moon lies at one of the two points where the planes intersect. If this were not so, there would be lunar eclipses with every full moon and solar eclipses with every new moon.

When the moon does pass directly into Earth's shadow, a circular darkening gradually advances across the moon's face, totally covering it within about an hour. Usually the moon remains dimly visible as sunlight passes through and is refracted (bent) by Earth's atmosphere, thus reaching the otherwise darkened lunar surface. After another hour or two, the moon has left the shadow and again appears full. Interestingly, during the partial phases of a lunar eclipse, Earth's shadow is easily seen to be circular. This indicated to at least some early astronomers that Earth is approximately spherical.

When the moon's shadow falls on Earth, a much more dramatic spectacle occurs. The shadow consists of two parts—the umbra and the penumbra. In the penumbra, the moon blocks only part of the sun, and on Earth many people may not notice anything unusual. The umbra, however, is the cone shaped region in which the sun's light is totally blocked. When the tip of this shadow reaches Earth, the moon's disk appears big enough in Earth's sky to cover the sun. This patch of darkness is rarely more than 150 miles (240 kilometers) wide. It races across Earth at over 1,000 miles (1,600 kilometers) per hour as the moon moves. Those in its path see the sun completely disappear from the sky and are enveloped in darkness almost as deep as night for up to about 7 minutes.

During totality the sun's corona, or outer atmosphere, can be seen surrounding the black silhouette of the moon's disk. The corona is only about as bright as a full moon and is normally blotted out by the bright daytime sky. An eclipse provides a rare opportunity to see the corona.

Sometimes the umbra fails to reach Earth's surface, meaning that the moon is too far from Earth to appear big enough to totally cover the sun. This leaves a thin but bright ring of sunlight at mid-eclipse. Such eclipses are called annular. They occur a bit more frequently than total eclipses do.

Only the total phase of a solar eclipse is safe to view, as looking at even a small part of the sun can cause permanent eye damage. Various filters and other methods exist to allow safe viewing of partial phases, but even these should be used with care.

Rocks from Outer Space

Sometimes one can see a flash of light streak across the night sky and disappear. Although this is commonly called a shooting star, real stars do not shoot through the sky any more than the sun does.

As the moon revolves in an almost circular path around Earth, Earth moves in a similar path around the sun. Both motions combine to give the moon a wavy orbit.

Patrick Pleul-dpa/Landov

A glowing streamer of the northern lights appears in the sky above a lake near Kautokeino, Norway.

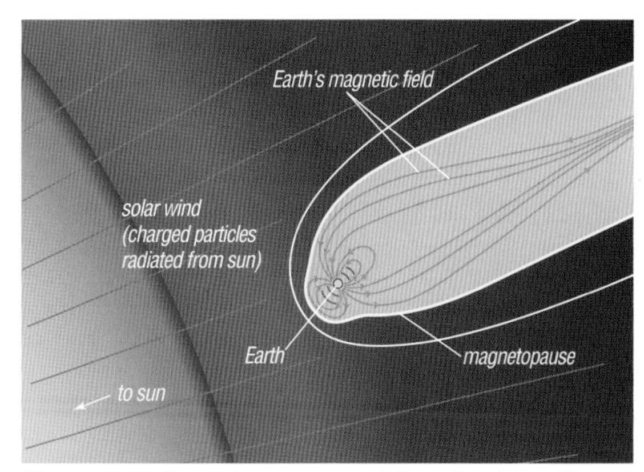

The sun gives off a continuous stream of charged particles. When this stream, called the solar wind, reaches Earth, it deforms Earth's magnetic field. Some of the particles spiral down near the magnetic poles, where they cause auroras.

Many small chunks of stone, metal, or other materials orbit the sun. Sometimes they enter Earth's atmosphere, and the friction generated by their great speed causes them to burn up. The fragments may either vaporize before traveling far or actually hit the ground.

These objects have different names depending on their location. One that is beyond Earth's atmosphere is called a meteoroid. A meteoroid that enters Earth's atmosphere is called a meteor. A meteor that actually lands on Earth's surface is called a meteorite.

Meteorites, which are sturdy enough to reach the ground, apparently are pieces of asteroids. Asteroids are huge rocks that orbit the sun. Most meteors that burn up in the atmosphere are tiny dustlike particles, the remains of disintegrated comets. Comets are flimsy objects made mostly of frozen water, frozen gases, and some gritty material. They also orbit the sun.

Sometimes a swarm of meteoroids enters Earth's atmosphere, causing a meteor shower, with tens or hundreds of "shooting stars" flashing across the sky in less than an hour. Virtually all these meteors burn up in the upper atmosphere. A significant amount of dust and ash from meteors settles on Earth each day. The Leonid meteors caused the greatest meteor showers on record, in 1833 and 1966. These meteors appear every November, with especially dazzling displays about every 33 years. The Leonid meteors are so named because their motion relative to Earth makes them appear to come from the direction of the constellation Leo.

The Northern and Southern Lights

People who are relatively near the North or South Pole may see one of nature's most lavish displays—the aurora borealis (northern lights) or the aurora australis (southern lights). High in the skies over Earth's magnetic poles, electrically charged particles from the sun swarm down into Earth's atmosphere. As these particles collide with air molecules, brilliant sheets, streamers, or beams of colored lights are given off at heights ranging from about 50 to 200 miles (80 to 320 kilometers) up in Earth's atmosphere.

The streams of charged particles are known as the solar wind. The sun continually sends a flow of these particles out into space. During periods when the sun is unusually active—that is, when it has large sunspots on its surface—the solar wind is particularly strong. Huge swarms of the particles then reach Earth's atmosphere, causing large and brilliant auroras.

TOOLS AND TECHNIQUES OF ASTRONOMY

Astronomers are at a distinct disadvantage compared with practitioners of other sciences; with few exceptions, they cannot experiment on the objects they study. Virtually all the information available is in the form of electromagnetic radiation (such as light) arriving from distant objects. Fortunately, this radiation contains an amazing number of clues to the nature of the objects emitting it.

The primary mirror of the Keck I telescope measures some 33 feet (10 meters) in diameter and is composed of 36 hexagonal segments. This huge mirror allows astronomers to view objects that are millions of times fainter than could be seen with the naked eye.

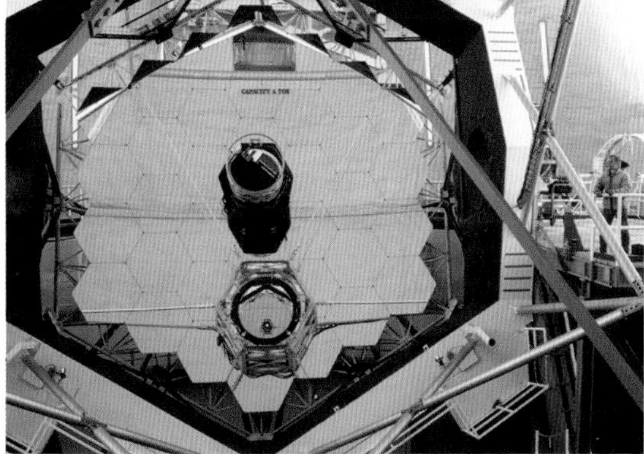

Roger Ressmeyer/Corbis

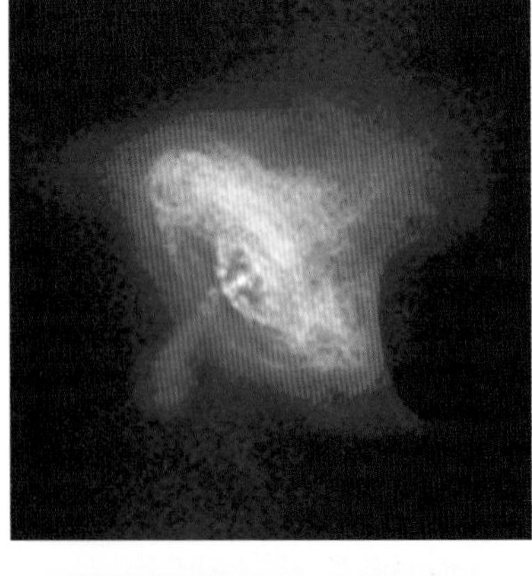

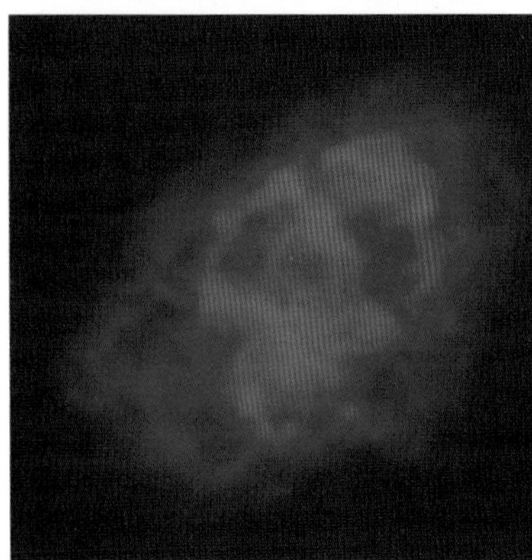

Four images of the Crab nebula captured at different wavelengths of electromagnetic radiation reveal different features. The nebula is the remains of a star that Chinese astronomers saw explode in AD 1054. At its center is a pulsar, or the star's very dense collapsed core that spins rapidly while beaming out radiation. The Crab nebula is still undergoing violent expansion. The X-ray image from the Chandra X-Ray Observatory (top left) reveals high-energy particles that the pulsar seems to have blasted outward, in rings from the center and in jets perpendicular to the rings. Over time, the particles move farther outward and lose energy to radiation. The cloud of lower-energy gas and dust surrounding the pulsar can be seen in the images taken at longer wavelengths: in visible light (top right), in infrared (bottom right), and in radio waves (bottom left). The image in visible light shows filaments of hot gas rushing outward. The radio image shows the lowest-energy particles and thus the largest extent of the expanding nebula. (These images are not to scale. The area of the nebula shown in visible light is actually 60 percent larger than the area shown in X-rays. The area shown in radio waves is about 20 percent larger than that in visible light.)

(Top left) NASA/CXC/SAO; (top right) Palomar Observatory; (bottom right) 2MASS/UMass/IPAC-Caltech/NASA/NSF; (bottom left) VLA/NRAO

Electromagnetic radiation travels in the form of waves, or oscillating electric and magnetic fields. In its interaction with matter, however, it is best understood as consisting of particles, called photons. These waves occur in a vast variety of frequencies and wavelengths. In order of increasing frequency (decreasing wavelength) these parts of the electromagnetic spectrum are called radio waves, microwaves, infrared, visible light, ultraviolet, X-rays, and gamma rays. As particles, radio wave photons carry the least amount of energy and gamma rays the most.

Telescopes

Naturally, the first part of the spectrum to be studied with instruments was visible light. Telescopes, first used for astronomy by Galileo in 1609, use lenses or mirrors to form images of distant objects. These images can be viewed directly or captured using film or electronic devices. Telescopes gather more light than the naked eye and magnify the image, allowing finer details to be seen. Even though early telescopes were crude by today's standards, they almost immediately allowed discoveries such as the moon's craters, Jupiter's moons, Saturn's rings, Venus' phases, sunspots, and thousands of previously unseen stars.

In the 20th century new technologies allowed the development of telescopes capable of detecting electromagnetic radiation all the way across the spectrum. Many objects emit most of their "light" at frequencies well outside the visible range. Even objects that do emit visible light often betray much more information when studied at other wavelengths.

By the 1990s optical (visible light) telescopes reached enormous size and power, a good example being the Keck telescopes on top of Mauna Kea in Hawaii. These two telescopes have collecting mirrors 33 feet (10 meters) in diameter, allowing detection of objects millions of times fainter than can be seen with the naked eye, with detail about a thousand times finer. Actually, astronomers seldom look through such telescopes directly. Instead, they use cameras to capture images photographically or newer, more sensitive detectors to capture images electronically. Most work is now done with electronic detectors, including charge-coupled devices (CCDs).

Since the 1940s radio telescopes have made great contributions. The largest single antenna, with a dish diameter of 1,000 feet (300 meters), is the Arecibo instrument in Puerto Rico. Huge arrays of multiple telescopes, such as the Very Large Array (VLA) in New Mexico, allow highly detailed imaging using radio waves, which otherwise yield rather "blurry" images. The largest is the Very Long Baseline Array (VLBA), consisting of ten dishes scattered over an area thousands of miles across. Data from these instruments are correlated using a technique called interferometry. The level of detail that can then be seen in radio-emitting objects (such as the centers of distant galaxies) is equivalent to discerning a dime at a distance of a few thousand miles.

A tremendous advance has been the placement of astronomical instruments in space. Telescopes and other instruments aboard unmanned spacecraft have explored all the sun's planets at close range. At least as important, though, have been large telescopes placed in Earth orbit, above the obscuring and blurring effects of Earth's atmosphere.

The best known of these telescopes is NASA's Hubble Space Telescope, which was launched in 1990 into an orbit 380 miles (610 kilometers) above Earth's surface. It initially returned disappointing images, owing to a mistake in the grinding of its 94.5-inch (2.4-meter) primary mirror. In 1993 space shuttle astronauts installed corrective optics, and ever since it has returned magnificent data. While Hubble is smaller than many groundbased telescopes, the lack of air to distort the images has generally allowed it better views than can be had from the ground, leading to many discoveries. Interestingly, a technology called adaptive optics now allows many ground-based telescopes to rival Hubble's level of detail, by removing much of the blurring effect of the atmosphere.

Less well known than Hubble but perhaps just as important are several other space telescopes that specialize in other parts of the spectrum. NASA's Compton Gamma Ray Observatory (whose mission lasted from 1991 to 2000) and Chandra X-ray Observatory (launched in 1999) have sent back a flood of data about objects such as neutron stars and black holes. These objects produce high-energy radiation that is largely blocked by Earth's atmosphere. NASA's Spitzer Space Telescope (launched in 2003) detects a

The Hubble Heritage Team (AURA/STScI/NASA)

A star's color indicates its surface temperature. The Hubble Space Telescope captured this dazzling image of a star cloud in the constellation Sagittarius. Most of these stars are fairly faint and orange or red, which is how the sun would appear. The blue and green stars are hotter than the sun, while the bright-red stars are red giants, which are much cooler stars near the end of their lives. The sun will eventually become a red giant.

wide range of infrared radiation, which is emitted by cooler objects, including interstellar clouds of gas and dust, where stars and planets form.

Spectroscopy: What Light Tells Astronomers

Stars give off a whole range of electromagnetic radiation. The kind of radiation is related to the temperature of the star: the higher the temperature of the star, the more energy it gives off and the more this energy is concentrated in high-frequency radiation. An instrument called a spectrograph can separate radiation into the different frequencies. The array of frequencies makes up the spectrum of the star.

The color of a star is also an indication of its temperature. Red light has less energy than blue light.

The absorption lines from a star or galaxy shift to longer wavelengths (red shift) when the object is receding from an observer. They shift to shorter wavelengths (blue shift) when the object is approaching an observer. When the absorption lines of an approaching star, a receding star, and the relatively stationary sun are shown against the background of a laboratory spectrum, it is clear that the lines occupy different positions on the spectrum. Austrian physicist Christian Doppler was the first to describe these shifts.

The Doppler Shift

The amount of shift depends on the velocity of the object in relationship to the observer: the greater the velocity, the greater the shift.

Absorption lines from an approaching object shift toward the violet (shorter wavelength).

Absorption lines from the sun are used for comparison.

Absorption lines from a receding object shift toward the red (longer wavelength).

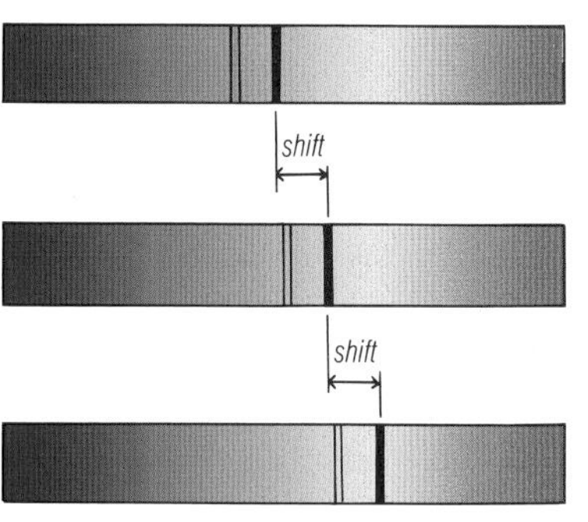

The colored bands are standard spectra. The black lines are absorption lines.

A reddish star must have a large amount of its energy in red light. A white or bluish star has a larger amount of higher-energy blue light, so it must be hotter than the reddish star.

Stars have bright or dark lines in their spectra. These bright or dark lines are narrow regions of extra-high emission or absorption of electromagnetic radiation. The presence of a certain chemical, such as hydrogen or calcium, in the star causes a particular set of lines in the star's spectrum. Since most of the lines found in stellar spectra have been identified with specific chemicals, astronomers can learn from a star's spectrum what chemicals it contains.

Spectrum lines are useful in another way, too. When an observer sees radiation coming from a source, such as a star, the frequency of the radiation is affected by the observer's motion toward or away from the source. This is called the Doppler effect. If the observer and the star are moving away from each other, the observer detects a shift to lower frequencies. If the star and the observer are approaching each other, the shift is to higher frequencies.

Astronomers know the normal spectrum-line frequencies for many chemicals. By comparing these known frequencies with those of the same set of lines in a star's spectrum, astronomers can tell how fast the star is moving toward or away from Earth.

Computer Modeling: Worlds Inside a Machine

While astronomers mostly cannot experiment with real astronomical objects in the laboratory, they can write computer programs to employ the laws of physics to simulate the structure and behavior of the actual objects. These models are never perfect, since both computing power and detailed knowledge of the structure and composition of the objects of interest are limited. In some situations, there are even uncertainties in the laws themselves. Nonetheless, these models can be adjusted until they closely match observable features and behavior of real objects.

Among the many types of astronomical phenomena that can be modeled are the evolution of stars, planetary systems, galaxies, and even the universe itself. Models of stars have successfully simulated their observed properties and supply predictions of what happens to them as they age. Other models have shown how planets can form from rotating clouds of gas and dust. Models of the early universe allow astronomers to study how large-scale structures such as galaxies developed as gravity accentuated tiny differences in the universe's density. As computers and modeling techniques have improved, this has become an ever more important tool of astronomy.

THE SOLAR SYSTEM

The solar system consists of the sun plus all the objects that orbit it. With more than 99 percent of the solar system's total mass and a diameter more than 100 times that of Earth and ten times that of Jupiter, the sun is quite naturally the center of the system. The spectrum, brightness, mass, size, and age of the sun and of nearby stars indicate that the sun is a typical star. Like most stars, the sun produces energy by thermonuclear processes that take place at its core. This energy maintains the conditions needed for life on Earth.

NASA

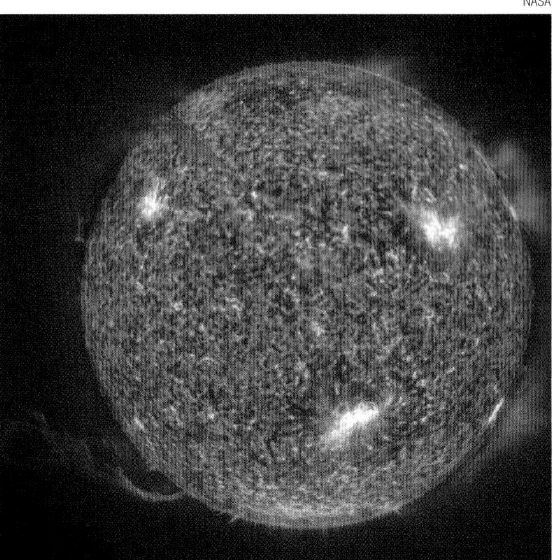

At the center of the solar system is the sun, which produces an enormous amount of energy. This image was taken in extreme ultraviolet light by the Earth-orbiting Solar and Heliospheric Observatory (SOHO) satellite. Nearly white areas are the hottest, while deep-red regions are the coolest. A massive prominence can be seen at lower left.

As has been mentioned above, Earth is not the only body to circle the sun. Many chunks of matter, some much larger than Earth and some microscopic, are caught in the sun's gravitational field. Eight of the largest of these chunks are called planets. Earth is the third planet from the sun. The smaller chunks of matter include dwarf planets, natural satellites

Kepler's Second Law

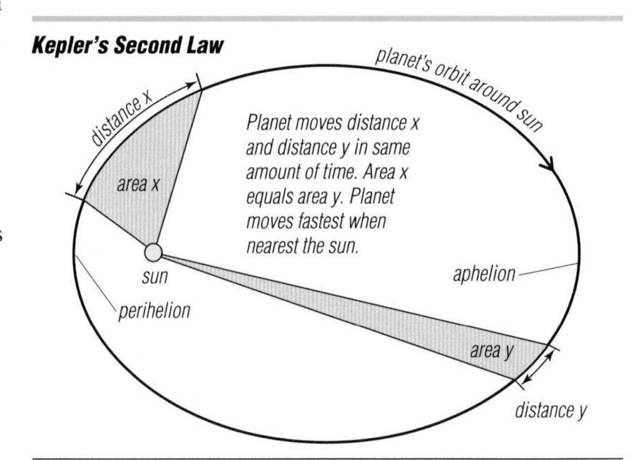

Kepler's second law of planetary motion describes the speed of a planet traveling in an eliptical orbit around the sun. It states that a line between the sun and the planet sweeps equal areas in equal times. Thus, the speed of the planet increases as it nears the sun and decreases as it recedes from the sun.

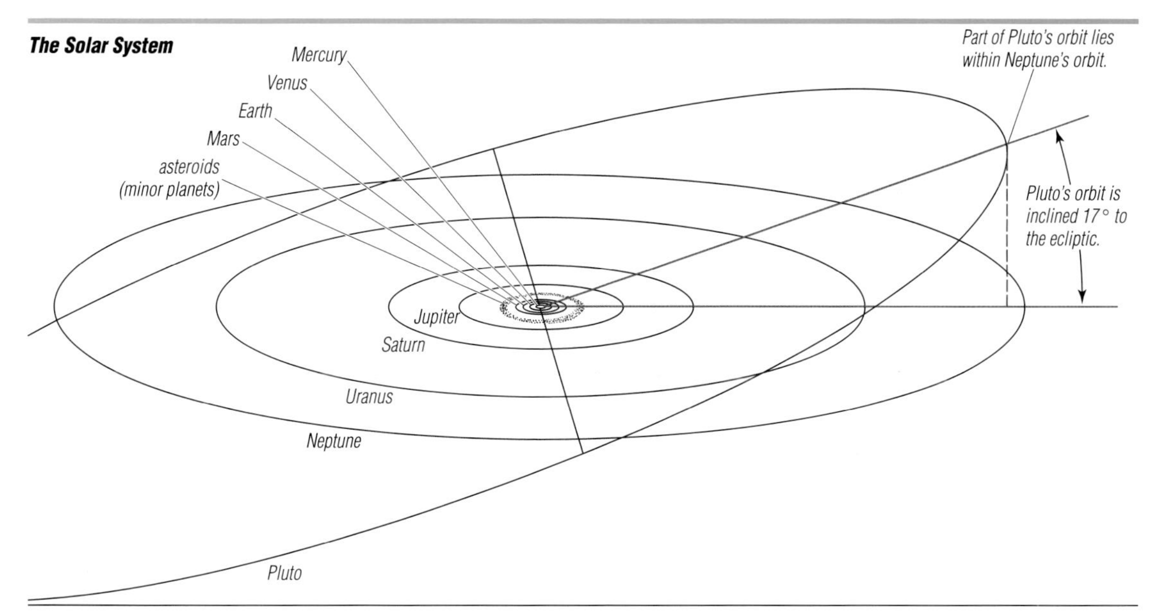

Many pieces of matter are held in the sun's enormous gravitational field. Together with the sun, they make up the solar system. The paths of the planets are huge ellipses with the sun at one focus. All the planets travel in the same direction around the sun. The dwarf planet Pluto's orbit is more elongated than those of the planets. It is also more tilted relative to the ecliptic plane, which is the plane of Earth's orbit.

(moons), asteroids, comets, meteoroids, and the molecules of interplanetary gases.

Kepler's Laws of Planetary Motion

In the early 1600s astronomers were beginning to accept the idea that Earth and the planets revolve around the sun, rather than that the sun and the planets revolve around Earth. Astronomers were still unable, however, to describe the motions of the planets as accurately as they could measure them. The German astronomer Johannes Kepler was finally able to describe planetary motions using three mathematical expressions, which came to be known as Kepler's laws of planetary motion.

In carefully studying Mars, Kepler found that its orbit is not circular, as had been assumed. Rather, the orbits of the planets are elliptical, with the sun at one of two fixed points in the ellipse called foci. Also, as a planet travels around the sun, its speed is greater when it is closer to the sun. An imaginary line drawn from the moving planet to the sun would sweep out equal areas in equal time intervals. Finally, Kepler found a mathematical relationship between a planet's average distance from the sun and its orbital period (the time it takes to complete an orbit). Specifically, he found that the squares of the planets' orbital periods are proportional to the cubes of their average distances from the sun.

To find these laws, Kepler had to effectively make a scale drawing of the solar system. He did this using extremely accurate observations collected by his deceased former employer Tycho Brahe. Kepler used a relative distance scale in which the average distance from Earth to the sun was called one astronomical unit.

Kepler did not have a particularly accurate value for the astronomical unit. To help find this distance, later astronomers were able to use methods such as parallax, an apparent shift in an object's position due to a difference in the observer's position (discussed below). Even more advanced methods have determined that Earth's average distance from the sun is in fact 92,955,808 miles (149,597,870 kilometers).

Newton's Law of Universal Gravitation

Kepler's laws described the positions and motions of the planets with great accuracy, but they did not explain what caused the planets to follow those paths. If the planets were not acted on by some force, scientists reasoned, they would simply continue to move in a straight line past the sun and out toward the stars. Some force must be attracting them to the sun.

The English scientist Isaac Newton calculated that in order for Kepler's laws to have the form they do, this force must grow weaker with increasing distance from the sun, in a particular way called an inverse square law. He also realized that the moon's curved path around Earth was a type of acceleration toward Earth. He calculated this acceleration to be much less than that of an apple falling from a tree. In comparing these accelerations, he found their difference to be described by the same inverse square law that described the force the sun exerted on the planets. Even the orbits of the other planets' moons could be similarly explained. Newton concluded that all

NASA/Lunar and Planetary Laboratory

The planets of the solar system are shown in a montage of images scaled to show the planets' approximate sizes relative to one another. The yellow segment at left represents the sun, to scale. The planets, from left to right, are Mercury, Venus, Earth, Mars, Jupiter, Saturn, Uranus, and Neptune.

masses in the universe attract each other with this universal force, which he called gravitation.

The Planets

Up to the 18th century people knew of seven bodies, besides Earth, that moved against the background of the fixed stars. These were the sun, the moon, and the five planets that are easily visible to the unaided eye: Mercury, Venus, Mars, Jupiter, and Saturn. Then, in 1781, William Herschel, a German-born English organist and amateur astronomer, discovered a new planet, which became known as Uranus.

Uranus' motion did not follow the exact path predicted by Newton's theory of gravitation. This problem was happily resolved by the discovery of an eighth planet, which was named Neptune. Two mathematicians, John Couch Adams and Urbain-Jean-Joseph Le Verrier, had calculated Neptune's probable location, but it was the German astronomer Johann Gottfried Galle who located the planet, in 1846.

Even then some small deviations seemed to remain in the orbits of both planets. This led to the search for yet another planet, based on calculations made by the U.S. astronomer Percival Lowell. In 1930 the U.S. astronomer Clyde W. Tombaugh discovered the object that became known as Pluto.

Pluto is an icy body that is smaller than Earth's moon. The mass of Pluto has proved so small—about $^{1}/_{500}$ of Earth's mass—that it could not have been responsible for the deviations in the observed paths of Uranus and Neptune. The orbital deviations, however, had been predicted on the basis of the best estimates of the planets' mass available at that time. When astronomers recalculated using more accurate measurements taken by NASA's Voyager 2 spacecraft in 1989, the deviations "disappeared."

For some 75 years astronomers considered Pluto to be the solar system's ninth planet. This tiny distant body was found to be unusual for a planet, however, in its orbit, composition, size, and other properties. In the late 20th century astronomers discovered a group of numerous small icy bodies that orbit the sun from beyond Neptune in a nearly flat ring called the Kuiper belt (discussed in full below). Many of Pluto's characteristics seem similar to those of Kuiper belt objects, and several of those objects are about the same size as Pluto. Many astronomers began to consider Pluto one of the larger members of the Kuiper belt.

In 2006 the International Astronomical Union, the organization that approves the names of celestial objects, removed Pluto from the list of planets. Instead, it made Pluto the prototype of a new category of objects, called dwarf planets. Like planets, dwarf planets are large, nearly spherical objects that orbit the sun. Under the 2006 definition, however, a planet must be massive enough for its gravity to have cleared away rocky and icy debris from the area around its orbit. Many small Kuiper belt objects lie in the orbital vicinity of Pluto, so it is not a planet according to the IAU's definition. Along with Pluto, the first objects that the IAU designated as dwarf planets were Eris, a Kuiper belt object that is slightly larger than Pluto, and Ceres, the largest asteroid.

All eight planets travel around the sun in elliptical orbits that are close to being circles. Mercury has the most eccentric (least circular) orbit. All the planets travel in one direction around the sun, the same direction in which the sun rotates. Furthermore, all the planetary orbits lie in very nearly the same plane. Mercury's is the most tilted, being inclined about 7 degrees relative to the plane of Earth's orbit (the ecliptic plane). By comparison, Pluto's orbit is tilted about 17 degrees from the ecliptic plane. Its orbit is also more eccentric than Mercury's.

Except for Venus and Uranus, each planet rotates on its axis in a west-to-east motion. In most cases the spin axis is nearly at a right angle to the plane of the planet's orbit. Uranus, however, is tilted so that its spin axis lies almost in its plane of orbit.

The planets can be divided into two groups. The inner planets—Mercury, Venus, Earth, and Mars—lie between the sun and the asteroid belt. They are dense, rocky, and small. Since Earth is a typical inner planet, this group is sometimes called the terrestrial, or Earth-like, planets.

The outer planets—Jupiter, Saturn, Uranus, and Neptune—lie beyond the asteroid belt. They are also called the Jovian, or Jupiter-like, planets. These are much larger and more massive than the inner planets. Jupiter has 318 times Earth's mass and in fact is more massive than all the other planets combined. Being made mostly of hydrogen and helium (mainly in liquid forms), the Jovian planets are also much less dense than the inner planets.

Natural Satellites of the Planets

Six of the planets—Earth, Mars, Jupiter, Saturn, Uranus, and Neptune—are known to have satellites. Because the moon is large in comparison with Earth, the Earth-moon system is sometimes called a double planet. Dwarf planets and asteroids can also have moons. Pluto's large satellite, Charon, has just over half the diameter of Pluto. Although several other satellites are much larger than either Earth's moon or Charon, these other satellites are much tinier, by comparison, than the bodies they circle.

Many of the natural satellites are fascinating worlds in their own right. Jupiter's moon Io has numerous active volcanoes spewing sulfur compounds across its surface. Europa, Jupiter's next moon out, may well have a vast ocean of liquid water underneath its icy crust. Neptune's Triton has mysterious geysers erupting in spite of frigid surface temperatures near –400° F (–240° C).

Also of great interest are Saturn's moons, especially Titan and Enceladus. Titan, its largest moon, has a thick, cold, hazy atmosphere of nitrogen and methane. On its surface, drainage channels—apparently carved by showers of methane rain—cut through a crust of water ice and empty into flat areas, which may be dune fields, methane mudflats, or perhaps liquid methane lakes. Although Enceladus is small and very cold, it is geologically active, with geysers near the south pole that spout water vapor and water ice.

Asteroids

On Jan. 1, 1801, the Italian astronomer Giuseppi Piazzi found a small planetlike object in the large gap between the orbits of Mars and Jupiter. This rocky object, later named Ceres, was the first and largest of thousands of asteroids, or minor planets, that have been discovered. While most asteroids are found in a belt between Mars and Jupiter, there are a few others. Some cross Earth's orbit and may present the threat of a rare collision with Earth at some time in the future.

Comets, the Oort Cloud, and the Kuiper Belt

Comets are among the most unusual and unpredictable objects in the solar system. They are small bodies composed mostly of frozen water and gases, with some silicate grit. This composition and the nature of their orbits suggest that comets were formed before or at about the same time as the rest of the solar system.

Photo NASA/JPL/Caltech (NASA photo # PIA00136)

Asteroid Ida and its tiny moon, Dactyl, are shown in an image taken by the Galileo spacecraft in 1993. Ida is a main-belt asteroid about 35 miles (56 kilometers) in length. It shows the irregular shape and impact craters characteristic of many asteroids.

Comets apparently originate beyond the orbit of Neptune. At such distances from the sun, they maintain very low temperatures, preserving their frozen state. They become easily visible from Earth only if they pass close to the sun. As a comet approaches the sun, some of its ices evaporate. The solar wind pushes these evaporated gases away from the head of the comet and away from the sun. This temporarily gives the comet one or more long glowing tails that point away from the sun.

Determining the source of comets has been a puzzle for astronomers. Some comets return to the inner solar system periodically, traveling in long, elliptical orbits that may reach from Earth's orbit to beyond Neptune. Halley's comet, for example, appears about every 76 years. Comets lose material with each pass near the sun, however, and can probably survive only a few hundred such visits before their volatile materials are exhausted. This means that they could have traveled on such orbits for only a small fraction of the solar system's widely accepted 4.6-billion-year history.

Other comets' orbits have been traced out to tens of thousands of astronomical units and have periods of millions of years. Some of these comets may in fact be making their first ever visits to the inner solar system. Such considerations led Jan Oort in 1950 to suggest the existence of a vast, spherical cloud, containing perhaps billions of comets. Disturbances such as the gravitational influence of passing stars could deflect these comets toward the sun.

Gerard P. Kuiper proposed in 1951 that another group of icy bodies, including dormant comets, might exist in a belt just outside Neptune's orbit. Discoveries starting in the 1990s have confirmed Kuiper's hypothesis, as hundreds of objects have been found at about the distance he predicted. The belt is thought to contain many millions of icy objects, most of them small. However, the largest Kuiper belt objects, including Eris and Pluto, are large enough to be considered dwarf planets.

Current thinking suggests that many of the short-period comets, or those that complete an orbit in less than 200 years, may have originated in the Kuiper

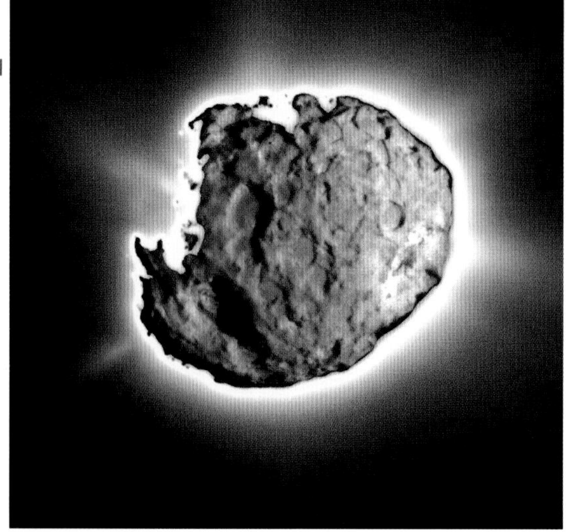

NASA/JPL-Caltech

The Stardust spacecraft took this composite image of comet Wild 2's icy nucleus during a flyby in 2004. It combines a short-exposure image that resolved surface detail and a long-exposure image that captured jets of gas and dust streaming away into space.

belt. They were perhaps directed into the inner solar system by collisions with each other and gravitational encounters with Neptune. Long-period comets are thought to originate in the Oort cloud (whose existence is considered highly probable, but not proven). The cloud may have been produced long ago, as icy bodies near and inside Neptune's orbit were thrown far out from the sun by gravitational encounters with the outer planets.

The Origin and Future of the Solar System

The most widely accepted model for the origin of the solar system combines theories elaborated by Kuiper and Thomas Chrowder Chamberlin. Astronomers believe that about 4.6 billion years ago, one of the many dense globules of gas and dust clouds that exist in the galaxy contracted into a slowly rotating disk called the solar nebula. The hot, dense center of the disk became the sun. The remaining outer material cooled into small particles of rock and metal that collided and stuck together, gradually growing into larger bodies to become the planets and their satellites.

In the cold outer parts of the new solar system, some of these bodies collected large amounts of hydrogen and helium from the solar nebula, thus becoming the "gas giants"—Jupiter, Saturn, Uranus, and Neptune. Closer to the sun, these light elements were mostly driven off by the higher temperatures and particles streaming off the sun. Smaller, rocky planets—Mercury, Venus, Earth, and Mars—developed there. Uncollected debris became asteroids and (in the outer regions) comets.

The sun is slowly getting brighter as it consumes its reservoir of hydrogen and turns this into helium. If current computations of stellar evolution are correct, the sun will grow much brighter and larger in about 5 billion years, making Earth much too hot for life to endure. Later the sun will have exhausted its nuclear energy source and will begin to cool. In the end it will become a white dwarf star, with all its matter packed densely into a space not much bigger than Earth. Around it will orbit frozen wastelands, the planets that survived the solar upheavals.

Does Life Exist Elsewhere?

Life as we know it, particularly in its higher forms, can exist only under certain chemical and physical conditions. The requirements for life are not fully known, but they almost surely include a reasonable temperature range, so that chemical bonding can occur, and a source of energy, such as sunlight or heat coming from the interior of a planet. It has also been commonly assumed that a solvent such as water and some protection from ultraviolet radiation are needed. A number of environments within the solar system may meet these criteria. For example, organisms could exist in the subsurface permafrost of Mars or in an ocean under the icy crust of Jupiter's moon Europa. Some comets and asteroids contain organic matter (meaning carbon-based molecules, not necessarily resulting from life). This suggests that the basic ingredients for life are common in the solar system.

Mars is an intriguing place to look for life. Spacecraft have photographed large features that appear to be dry riverbeds. Data from NASA's Spirit and Opportunity rovers in the early 2000s strongly suggest that liquid water once existed on the planet's surface. Also, data from the European Space Agency's Mars Express orbiter and from Earth-based telescopes suggest that small amounts of methane are being released from beneath the surface in places, and a possible source for this could be subsurface colonies of bacteria.

In 1976 the Viking landers looked for evidence of life in the Martian soil. They found no organic molecules. However, a couple of Viking experiments that looked for signs of metabolic processes, such as the labeled-release experiment, yielded seemingly positive results. These findings have been widely (but inconclusively) interpreted as a result of strange chemical reactions rather than life. While life has not been found on Mars, many scientists think that it may have existed in a wetter past, and some believe it may have survived into the present.

Discoveries of life existing in extreme or unusual environments on Earth—such as in hot bedrock miles beneath the surface and in colonies near volcanic vents on the deep sea floor—have widened prospects for finding life elsewhere. No place in the solar system other than Earth, however, is easily suitable for human colonization or for large land plants or animals. It is possible that other stars may be orbited by more Earth-like planets. In fact, the number of such worlds in the universe may be truly enormous. However, the only place life has been found so far is on Earth. One example is very little to go on, especially if we are part of the example. With little information regarding the likelihood of life arising in other places, even under Earth-like conditions, discussion of life elsewhere remains speculative.

THE STARS

Looking at the night sky through a telescope, or even with the naked eye, one can see a complex display. Different cultures around the world have imagined different patterns in the way the stars appear. Constellations are groups of stars that seem to form the shapes of people, animals, or objects. The first step in finding one's way among the stars is usually to learn to recognize a few basic constellations, such as the Big Dipper and Orion.

The stars in constellations are not necessarily close to each other in space. For example, though the middle five stars of the Big Dipper are relatively close together, the first and last stars only seem to be in the same group. They are actually much farther from Earth than the other five, and they are even slowly moving in different directions. Some parts of Orion are relatively close together, but Betelgeuse (pronounced "beetle juice"), the bright red star at the top, is much nearer to Earth.

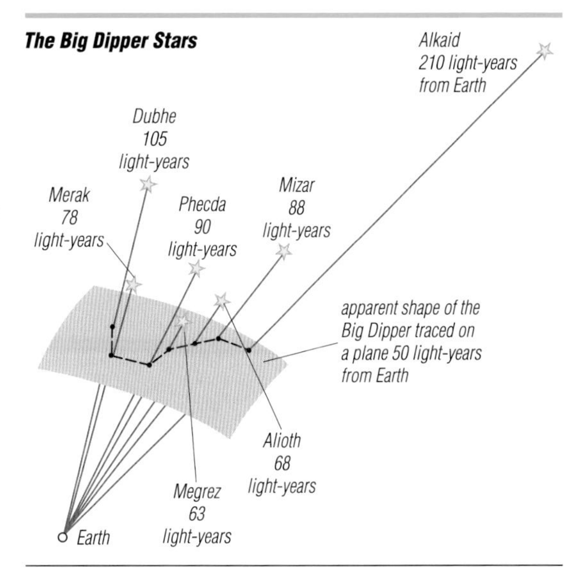

Although the stars of the Big Dipper seem to belong together, they are actually widely separated. A person looking at the Big Dipper stars from a position in space different from that of Earth would see them in a different shape, or they might seem completely unrelated to each other in the sky.

Coordinate Systems

Astronomers need to record the exact locations of stars. Within limits, it is useful to locate objects within constellations. Numerical coordinate systems are used to record the locations of celestial objects more precisely. These systems are like the coordinate system of latitude and longitude used for Earth.

Different celestial coordinate systems have been devised. To be useful they must take into account that Earth has two regular motions in relation to the stars. Its rotation causes the sphere of stars to appear to make a complete circle around the planet once a day. And Earth's revolution around the sun causes the apparent star positions at a particular hour to shift from day to day, so that they return to their "original" position after a year.

The horizon, or azimuth, system is based on Earth's north-south line and the observer's horizon. It uses two angles called azimuth and altitude. Azimuth locates the star relative to the north-south line, and altitude locates it relative to the plane of the horizon. For this system to be useful, the time of the observation and the location from which the observation was made must be accurately known.

The equator system is based on the concept of the celestial sphere. All the stars and other heavenly bodies can be imagined to be located on a huge sphere that surrounds Earth. The sphere has several imaginary lines and points. One such line is the celestial equator, which is the projection of Earth's equator onto the celestial sphere. Another is the line of the ecliptic, which is the sun's apparent yearly path along this sphere. The celestial equator and the ecliptic intersect at two points, called the vernal equinox and the autumnal equinox. (When the sun is at either point, day and night on Earth are equally long.) The north and south celestial poles are extensions of the North and South poles of Earth along Earth's axis of rotation.

In the equator system the position of a star is given by coordinates called declination and right ascension. The declination locates the star by its angular distance north or south of the celestial equator. The right

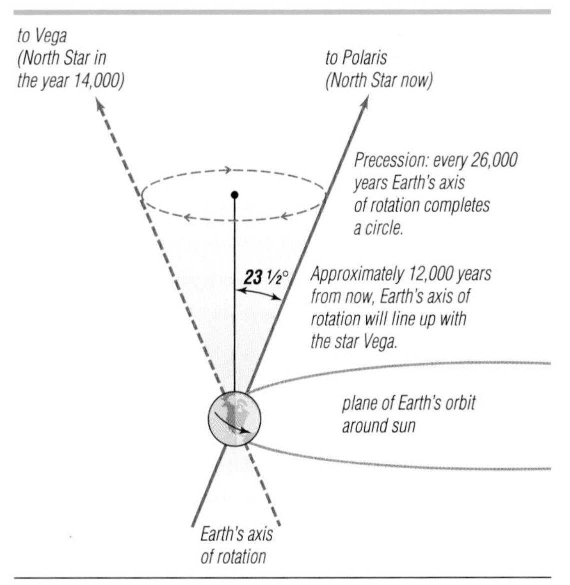

Earth's axis of rotation slowly wobbles in a toplike motion called precession. This means that the successive positions of the celestial north and south poles trace out large circles in the sky, completing the circles about every 26,000 years. The North Pole points toward different stars—and sometimes at no star in particular—as it travels in this circle. However, this precession is so slow that a person would not notice it in a lifetime.

Apparent Motion of Stars

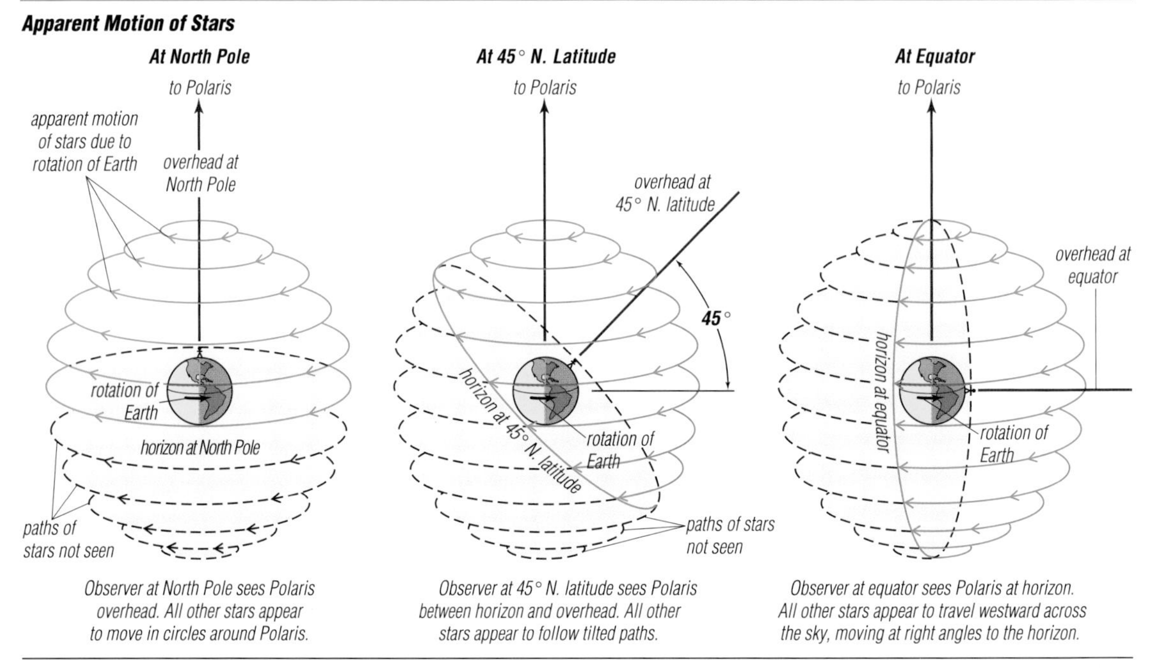

A person standing on Earth does not feel the motion of the planet as it spins. The sun and the stars appear to follow paths that lead from east to west across the sky.

ascension locates the star by its angular distance east or west of the vernal equinox. Since this system is attached to the celestial sphere, all points on Earth (except the poles) are continually changing their positions under the coordinate system.

Determining the Distance to Stars

Fixing stars on an imaginary sphere is useful for finding them from Earth, but it does not reveal their actual locations. One way to measure the distances of nearby stars from Earth is the parallax method.

For parallax measurements of stars, scientists make use of Earth's yearly motion around the sun. Because of this motion, observers on Earth view the stars from different positions at different times of the year. At any given time of year, Earth is 186 million miles (300 million kilometers) away on the opposite side of the sun from where it was six months before. Two photographs of a near star taken through a large telescope six months apart will show that the star appears to shift against the background of more distant stars. If this shift is large enough to be measured, astronomers can calculate the distance to the star.

More than four centuries ago the phenomenon of parallax was used to counter Nicolaus Copernicus' suggestion that Earth travels around the sun. Scientists of the time pointed out that if it did, stars should show an annual change in direction due to parallax. But, using the instruments available to them, they were unable to measure any parallax, so they concluded that Copernicus was wrong.

Astronomers now know that the stars are all at such tremendous distances from Earth that their parallax angles are extremely difficult to measure. Even modern instruments cannot measure the parallax of most stars.

Astronomers measure parallaxes of stars in seconds of arc. This is a tiny unit of measure; for example, a penny must be 2.5 miles (4 kilometers) away before it appears as small as one second of arc. Yet no star except the sun is close enough to have a parallax that large. Alpha Centauri, a member of the group of three stars nearest to the sun, has a parallax of about three quarters of a second of arc.

Astronomers have devised a unit of distance called the parsec—the distance at which the angle opposite the base of a triangle measures one second of arc when the base of the triangle is the radius of Earth's orbit around the sun. One parsec is equal to 19.2 trillion (19.2×10^{12}) miles (30.9 trillion kilometers). Alpha Centauri is about 1.3 parsecs distant.

Another unit used to record large astronomical distances is the light-year. This is the distance that light travels within a vacuum in one year—about 5.88 trillion miles (9.46 trillion kilometers). Proxima Centauri, part of the Alpha Centauri system, is the star closest to Earth (apart from the sun), yet it is about 4.3 light-years distant. Light takes more than four years to reach Earth from that distance.

Since parallax yields distances to only relatively nearby stars, other methods must be used for more distant ones. One of these methods is statistical parallax, in which the apparent motions across the sky

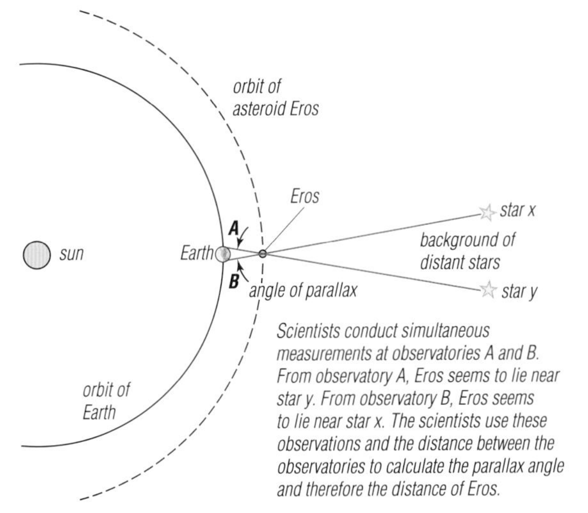

When the asteroid Eros approaches Earth, it frequently comes within 20 million miles and sometimes comes within 14 million miles of Earth. At such near approaches its distance is easy to measure by the parallax method.

of groups of stars are analyzed to determine their probable distance. Another method involves observing certain stars that vary regularly in brightness (discussed below).

Size and Brightness of Stars

Both the size and the temperature of a star determine how much radiation energy it gives off each second: this is the actual brightness of the star. It is also true, however, that the closer a star is to Earth, the more of its radiation energy will actually reach Earth and the brighter it will appear.

Astronomers express the brightness of a star in terms of its magnitude. Two values of magnitude describe a star. The apparent magnitude refers to how bright the star looks from Earth. The absolute magnitude of a star is the value its apparent magnitude would have if the star were ten parsecs from Earth. The apparent magnitude of a star depends on its size, temperature, and distance. The temperature is found from its spectrum; if the distance is known, then astronomers can calculate the size of the star and also assign a value for its absolute magnitude. The actual brightness of stars may be compared using their absolute magnitudes.

Certain stars whose brightness varies regularly provide an important way for astronomers to estimate the distances of remote galaxies. In such stars the actual brightness (absolute magnitude) is closely related to the period of their brightness variations. Astronomers can use the observed period to determine the actual brightness and then compare this with the apparent brightness to estimate the distance.

Astronomers have discovered all kinds of stars—from huge, brilliant red supergiants more than 100 times the sun's diameter to extremely dense neutron stars only about a dozen miles across. The sun lies in about the middle range of size and brightness of stars. The largest stars are the cool, reddish supergiants: they have low surface temperatures, but they are so bright that they must be extremely large to give off that much energy. White dwarf stars, on the other hand, are very faint in spite of their high surface temperatures and thus must be very small—only about the size of Earth.

What Is a Star?

Astronomers have found, using analysis of stars' spectra, that stars are made mostly of the simplest elements: hydrogen and helium. These elements are in the gaseous state. In most of the star, however, the temperature is so high (thousands to millions of degrees) that the gas is ionized (with electrons stripped away from the atomic nuclei)—a state called plasma.

The mutual gravitational attraction of a star's matter is what forces it into a roughly spherical shape. In fact, if there were nothing to counteract this inward force, the star would collapse to a very small size. The gravitational squeezing of the gas, however, heats it to very high temperatures. In the 1800s astronomers believed that this compression was actually the energy source for a star. This presented a problem. The sun could shine like this for only a few million years without shrinking so much that conditions on Earth would be greatly altered. Yet geological and biological evidence suggested that Earth has maintained the conditions for life for hundreds of millions of years.

The 20th century brought a solution to this problem. With the discovery of nuclear energy, astronomers could explain the sun's long-lasting power output as the result of nuclear fusion: hydrogen deep inside the sun was being fused together to form helium. This process is so energetic that it can counterbalance the inward force of gravity. Stars, then, are essentially

The Cat's Eye nebula is a spectacular example of what happens when a star of about the sun's mass dies. It is a so-called planetary nebula, which forms when a red giant ejects expanding shells of gas. Radiation from the hot central star makes the gaseous shells glow. The central star will eventually become a white dwarf.
NASA, ESA, HEIC, and The Hubble Heritage Team (STScI/AURA)

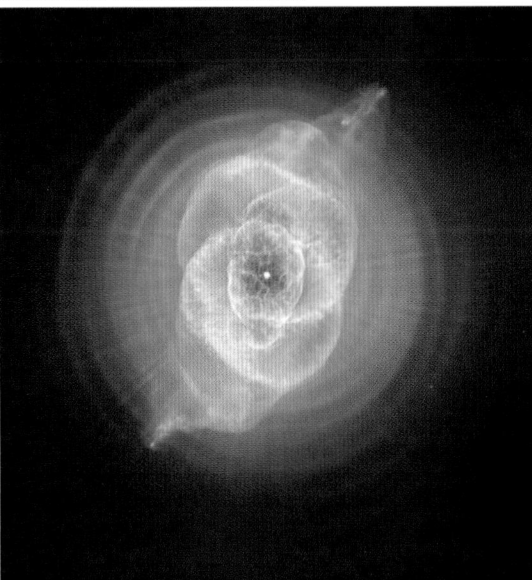

battlegrounds between two forces—the inward crush of gravity and the outward pressure from the heat generated by nuclear fusion.

The Lives of Stars

Stars are believed to form when large clouds of gas and dust, called nebulae, contract gravitationally (though other forces may also play a role). Eventually they become hot enough (several million degrees) in the center to start fusion of hydrogen into helium. By this time, the gas is glowing brightly, and a star is born.

This cannot last forever, though, as eventually most of the hydrogen "fuel" is converted into helium. In the largest stars, this takes only a few million years. Very-low-mass stars, with less gravitational pressure to battle, consume their fuel very slowly and may last a trillion years. The sun is intermediate, with an estimated lifetime of about 10 billion years, which it is believed to be almost halfway through.

When a star's core has been converted mainly into helium, dramatic changes occur in its structure. Computer models, backed up by observations of many stars at different stages, predict that stars like the sun will swell to about a hundred times their former diameter. After a relatively short period as such a red giant, the star will lose its outer layers, leaving a small, hot core. The core will then shrink to form a white dwarf star. Hundreds of such objects have been observed, generally confirming the predictions.

Stars born with much more mass than the sun undergo even more dramatic events. Under tremendous pressure, such a star performs numerous additional fusion reactions in its core, producing a wide range of elements, up to and including iron. At this point, the ultradense core can collapse suddenly, leading to a colossal explosion called a supernova. Many such events have been observed from Earth, some so bright that they were visible in broad daylight. For a few weeks the exploding star can outshine an entire galaxy of a hundred billion stars. The elements thrown out into space can become part of nebulae, eventually to be incorporated into future generations of stars and planets.

Neutron Stars and Black Holes

After some types of supernova explosions, an extremely dense core remains. This object, called a neutron star, is about the mass of the sun and is made mostly of neutrons. Its matter is so compact that a teaspoon of it has the mass of a small mountain. Some neutron stars spin rapidly while beaming radiation into space. If a beam intercepts Earth, astronomers may detect it as a series of pulses of radio waves or sometimes radiation at other wavelengths. Such a neutron star is referred to as a pulsar.

Even more massive stars may collapse to such high densities that their powerful gravitational pull will not allow even light, or anything else, to escape. They are called black holes. Black holes resulting from the collapse of a single, dying star may be only several miles across, but much larger ones—with the mass of millions of suns and the size of the solar system—are suspected to exist in the centers of many galaxies.

Walter Jaffe/Leiden Observatory, Holland Ford/JHU/STScI, and NASA

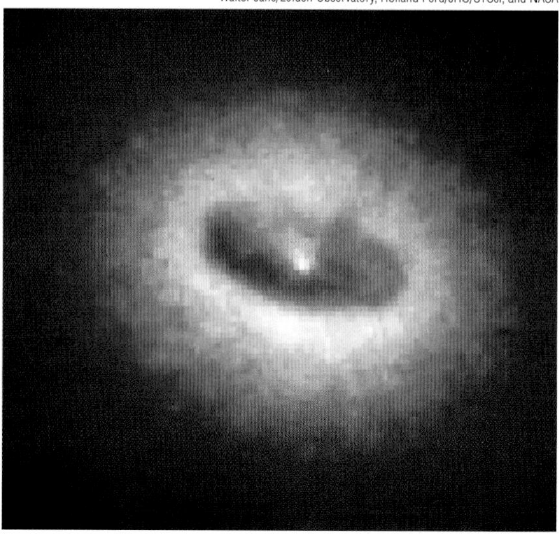

A Hubble Space Telescope image shows a disk of gas and dust that probably surrounds and feeds a huge black hole at its center. The enormous disk is some 300 light-years across. It lies at the center of a galaxy in the Virgo cluster. Astronomers think that black holes probably lie at the center of many galaxies.

Often, neutron stars and black holes are detectable only because of their effects on nearby companion stars. Gas (mainly hydrogen) is drawn off the companion star and then swirls rapidly down onto (or into) the neutron star or black hole. The violent compressional heating and acceleration of the gas causes it to emit X-rays, which can be detected from Earth-based satellites. Such double star systems are called X-ray binaries.

Planets of Other Stars

Astronomers have long thought that, like the sun, many or most stars should be accompanied by orbiting planets. These planets would be so distant from Earth, however, that their very faint light would be drowned out by the bright light of their "suns." It turns out that there are indirect methods of detecting such planets. An orbiting planet would cause a star to wobble slightly, and this wobble could be detected as alternating red and blue Doppler shifts of the star's light. Furthermore, the speed and period of the wobble could enable astronomers to estimate the planet's mass and distance from the star. This technique was first successfully used in 1995 to find a planet orbiting the star 51 Pegasi. During the next ten years, about 140 extrasolar planets were discovered in this way (plus a few by other means, such as the dip in light caused when a planet passes in front of a star).

Most of the planets found so far are at least as massive as Jupiter, yet they are closer to their stars than Mercury is to the sun. Such close-in, massive planets should be the easiest to detect, since they cause the greatest wobbles. But they are still a

VLT/ESO

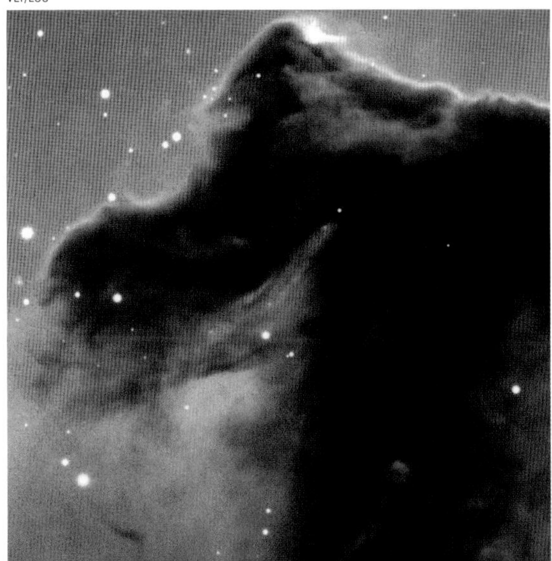

Dark nebulae, such as the Horsehead nebula in Orion, consist of clouds of interstellar dust, which scatter starlight. This composite image is from the European Southern Observatory. The brown areas are thick clouds of obscuring dust in the foreground, while the blue-green areas show scattered starlight. The red glowing areas result from hydrogen emitted by an H II region.

challenge to explain. Current theories of planet formation suggest that such large planets should form farther from the star, where temperatures are cold enough to allow collection of large amounts of gas. One possibility astronomers are considering is that these "hot Jupiters" formed farther out from their stars and migrated inward. This raises the question, though, of why our solar system has not experienced such planetary migration.

INTERSTELLAR MATTER

The space between the stars contains gas and dust at extremely low densities. This matter tends to clump into clouds. These clouds are called nebulae when they block more distant starlight, reflect starlight, or get heated by stars so that they glow. Interstellar dust is made of fine particles or grains. Although only a few of these grains are spread through a cubic mile of space, the distances between the stars are so great that the dust can block the light from distant stars. Many small, dark regions are known where few or no stars can be seen. These are dark nebulae, dust clouds of higher than average density that are thick enough to obscure the light beyond them.

Dust grains block blue light more than red light, so the color of a star can appear different if it is seen through much dust. To find the temperature of such a star, astronomers must estimate its color to be bluer than it appears because so much of its blue light is lost in the dust. When clouds of dust occur near bright stars they often reflect the starlight in all directions. Such clouds are known as reflection nebulae.

Interstellar gas is about 100 times denser than the dust but still has an extremely low density. The gas does not interfere with starlight passing through it, so it is usually difficult to detect. When a gas cloud occurs close to a hot star, however, the star's radiation causes the gas to glow. This forms a type of bright nebula known as an H II region. Away from hot stars interstellar gas is quite cool. Masses of this cool gas are called H I regions.

Interstellar gas, like most stars, consists mainly of the lightest element, hydrogen, with small amounts of helium and only traces of the other elements. The hydrogen readily glows in the hot H II regions. In the cool H I regions the hydrogen gives off radio-frequency radiation. Most interstellar gas can be located only by detecting these radio waves.

The hydrogen occurs partly as single atoms and partly as molecules (two hydrogen atoms joined together). Molecular hydrogen is even more difficult to detect than atomic hydrogen, but it must exist in abundance. Other molecules have been found in the interstellar gas because they give off low-frequency radiation. These molecules contain other atoms besides hydrogen: oxygen or carbon occurs in hydroxyl radicals (OH–) and in carbon monoxide (CO), formaldehyde (H_2CO), and many others, including many organic molecules.

Wherever there are large numbers of young stars, there are also large quantities of interstellar gas and dust. New stars are constantly being formed out of the gas and dust in regions where the clouds have high densities. Although many stars blow off part of their material back into the interstellar regions, the gas and dust are slowly being used up. Astronomers theorize that eventually a time will be reached when no new stars can be formed, and the star system will slowly fade as the stars burn out one by one.

THE GALAXIES

Stars are found in huge groups called galaxies. Scientists estimate that the larger galaxies may contain as many as a trillion stars, while the smallest may have fewer than a million. Large galaxies may be 100,000 or more light-years in diameter.

Galaxies may have any of four general shapes. Elliptical galaxies show little or no structure and vary in general shape from moderately flat and round or oval to spherical. Spiral galaxies have a small, bright central region, or nucleus, and arms that come out of the nucleus and wind around, trailing off like a giant pinwheel. In barred spiral galaxies, the arms extend sideways in a short straight line before turning off into the spiral shape. Both kinds of spiral systems are flat. Irregular galaxies are usually rather small and do not have a symmetrical shape.

Radio Galaxies, Quasars, and Dark Matter

Galaxies were long thought to be more or less passive objects, containing stars and interstellar gas and dust and shining by the radiation that their stars give off. When astronomers became able to make accurate observations of radio frequencies coming from space,

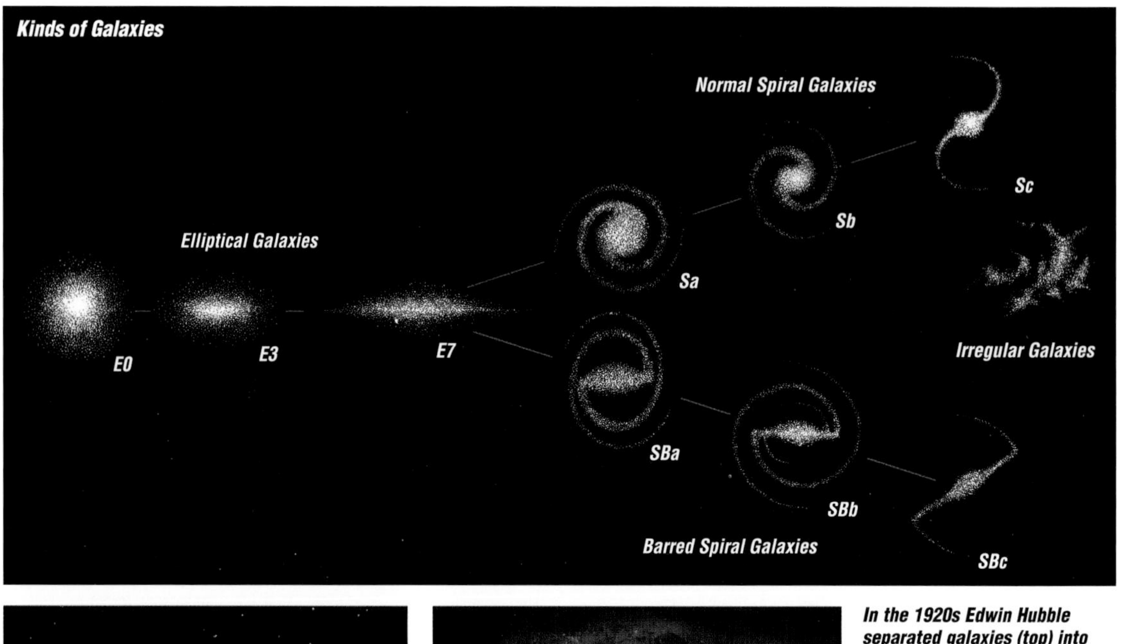

In the 1920s Edwin Hubble separated galaxies (top) into four general types according to their appearance and then classified each into subtypes. Four photos of specific galaxies illustrate this classification scheme. The elliptical galaxy (center left) is NGC 1316. This giant galaxy is unusual for one of the elliptical type in that it contains dust lanes. The spiral galaxy (center right) is the Whirlpool galaxy (M51). It is accompanied by a small, irregular companion galaxy, at the top right of that image. The galaxy NGC 1300 (bottom right) is an example of a barred spiral, while the Cigar galaxy (M82; bottom left) is irregular.
(Center left) VLT/ESO; (center right) NASA/ESA, S. Beckwith (STScI), and The Hubble Heritage Team (STScI/AURA); (bottom right and left) NASA, ESA, and The Hubble Heritage Team (STScI/AURA)

they were surprised to find that a number of galaxies emit large amounts of energy in the radio region. Ordinary stars are so hot that most of their energy is emitted in visible light, with little energy emitted at radio frequencies. Furthermore, astronomers were able to deduce that this radiation had been given off by charged particles of extremely high energy moving in magnetic fields.

How do such galaxies, called radio galaxies, manage to give so much energy to the charged particles and magnetic fields? Radio galaxies are also usually rather peculiar in appearance. Many galaxies, and the radio galaxies in particular, show evidence of interstellar matter expanding away from their centers, as though gigantic explosions had taken place in their nuclei. The giant elliptical galaxy known as M87 has a jet of material nearby that it apparently ejected in the past. The jet itself is the size of an ordinary galaxy.

Astronomers have found that, in many galaxies, stars near the center move very rapidly, apparently orbiting some very massive unseen object. The most likely explanation is that a giant black hole, with millions or even billions of times the sun's mass, lurks in the center of most large galaxies. As stars and gas spiral into these black holes, much of their mass vanishes from sight. The violent heating and compression produces a huge release of energy, including high-speed jets of matter (such as in M87).

Very distant galaxies are sometimes found to have extremely energetic sources of light and radio waves at their centers. These objects, called quasars, are generally believed to be several billion light-years from Earth. This means that astronomers who observe quasars are actually peering several billion years into the past. Most astronomers believe that quasars represent an early phase in the life of some galaxies, when the central black holes, with plenty of fresh gas and stars to consume, were generating huge amounts of energy.

Another problem has puzzled astronomers for years. Most, if not all, galaxies occur in clusters, presumably held together by the gravity of the cluster members. When the motions of the cluster members are measured, however, it is found in almost every case that the galaxies are moving too fast to be held together only by the gravity of the matter that is visible. Astronomers believe there must be a large amount of unseen matter in these clusters—perhaps ten times as much as can be seen. While some of this likely consists of objects such as black holes and neutron stars, most of it is believed to be "exotic dark matter," of unknown origin.

The Milky Way Galaxy

Like most stars, the sun belongs to a galaxy. Since the sun and Earth are embedded in the galaxy, it is difficult for astronomers to obtain an overall view of this galaxy. In fact, what can be seen of its structure is a faint band of stars called the Milky Way (the word galaxy comes from the Greek word for "milk"). Because of this, the galaxy has been named the Milky Way galaxy.

The visible band of the Milky Way seems to form a great circle around Earth. This indicates that the galaxy is fairly flat rather than spherical. (If it were spherical, the stars would not be concentrated in a single band.) The sun is located on the inner edge of a spiral arm. The center, or nucleus, of the galaxy is about 25,000 light-years distant, in the direction of the constellation Sagittarius. All the stars that are visible without a telescope belong to the Milky Way galaxy.

Not all the galaxy's stars are confined to the galactic plane. There are a few stars that occur far above or below the disk. They are usually very old stars, and they form what is called the halo of the galaxy. Evidently the galaxy was originally a roughly spherical mass of gas. Its gravity and rotation caused it to collapse into the disklike shape it has today. The stars that had been formed before the collapse

The name of our galaxy comes from the visual phenomenon of the Milky Way, a band of stars seen in Earth's night sky. This band is actually the major portion of the galaxy. Since Earth lies in the midst of the galaxy, the spiral structure is hidden by the nearest stars that lie in the plane of the galactic equator and form the Milky Way.

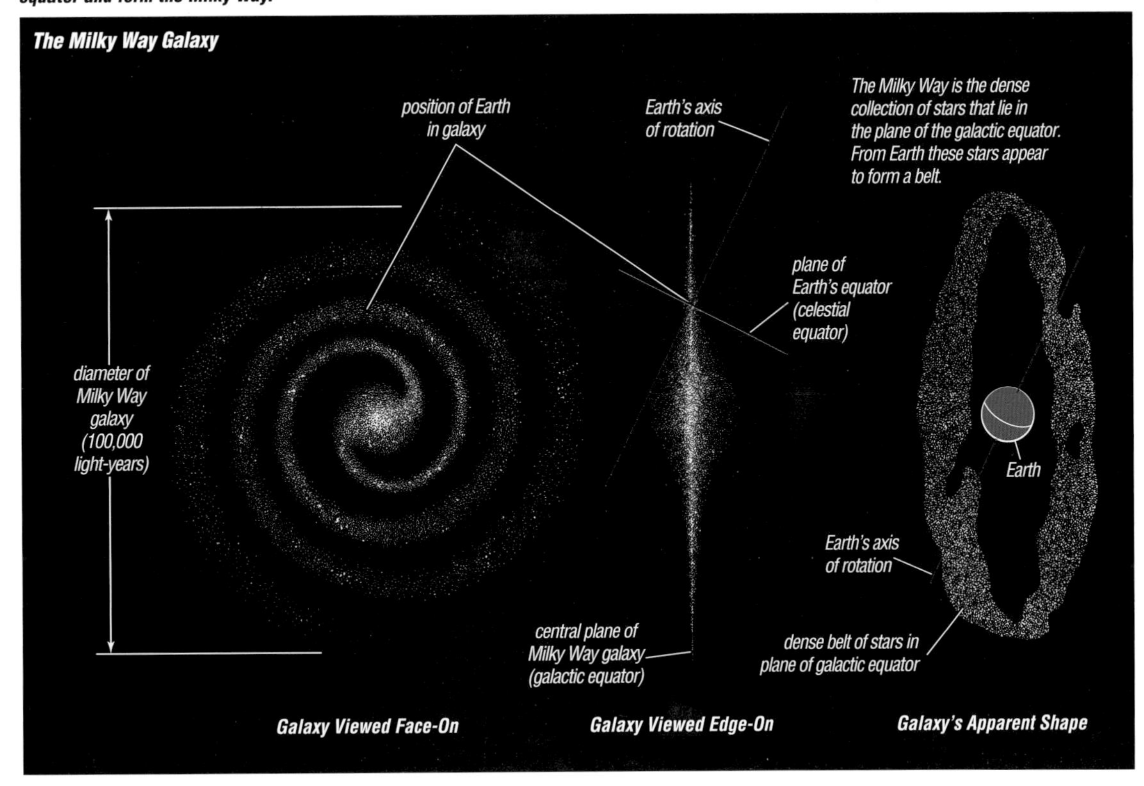

Evolutionary Theory of the Universe

Billions of Years Later

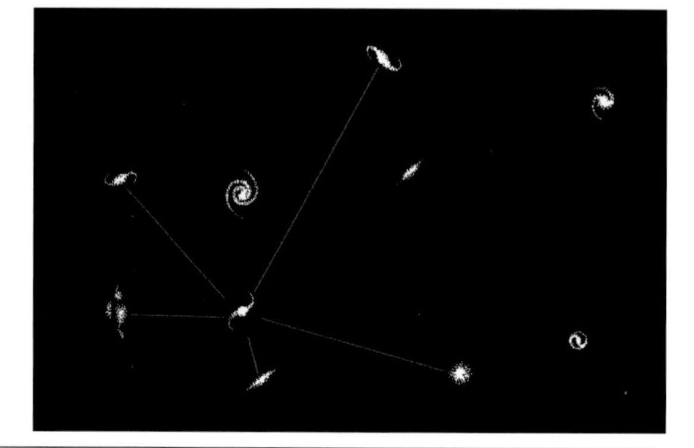

Now

According to the evolutionary, or big bang, theory of the universe, the universe is expanding while the total energy and matter it contains remains constant. Therefore, as the universe expands, the density of its energy and matter must grow progressively thinner. At left is a two-dimensional representation of the universe as it appears now, with galaxies occupying a typical section of space. At right, billions of years later the same amount of matter will fill a relatively larger volume of space.

remained in their old positions, but after the collapse further star formation could occur only in the flat disk.

All the stars in the galaxy move in orbits around its center. The sun takes about 200 million years to complete an orbit. The orbits of most of these stars are nearly circular and are nearly in the same direction. This gives a sense of rotation to the galaxy as a whole, even as the entire galaxy moves through space.

Dark clouds of dust almost completely obscure astronomers' view of the center of the Milky Way galaxy. Radio waves penetrate the dust, however, so radio telescopes can provide astronomers with a view of the galactic nucleus. In that region stars travel in very fast, tight orbits—which implies the existence of a huge mass at the center. In addition, the Earth-orbiting Chandra X-ray Observatory has detected flares of X-rays, lasting only a few minutes, in the region. Most astronomers believe these findings are best explained by the existence of a black hole—3 million times the sun's mass but only about a dozen times the sun's diameter—that is violently accelerating and compressing in-falling blobs of matter.

THE UNIVERSE

Cosmology is the scientific inquiry into the nature, history, development, and fate of the universe. By making assumptions that are not contradicted by the behavior of the observable universe, scientists build models, or theories, that attempt to describe the universe as a whole, including its origin and its future. They use each model until something is found that contradicts it. Then the model must be modified or discarded.

Cosmologists usually assume that, except for small irregularities, the universe has a similar appearance to all observers (and the laws of physics are identical), no matter where in the universe the observers are located or in which direction they look. This unproven concept is called the cosmological principle. One consequence of the cosmological principle is that the universe cannot have an edge, for an observer near the edge would have a different view from that of someone near the center. Thus space must be infinite and evenly filled with matter, or the geometry of space must be such that all observers see themselves as at the center. Also, astronomers believe that the only motion that can occur, except for small irregularities, is a uniform expansion or contraction of the universe.

Because the universe appears to be expanding, it seems that it must have been smaller in the past. This is the basis for evolutionary theories of the universe. If one could trace the galaxies back in time, one would find a time at which they were all close together. Observations of the expansion rate indicate that this was between 13 and 14 billion years ago. Thus emerges a picture of an evolving universe that started in some kind of "explosion"—the big bang. Some models of the universe have the expansion continuing forever. Others say that it will stop and be followed by a contraction back to a small volume again. However, data obtained since the late 1990s on the recession speeds of distant supernovas have strongly suggested that the expansion is actually accelerating. This may mean the universe will expand forever. Astronomers are currently trying to explain this acceleration. A current favorite explanation is the idea of dark energy, which might provide a repulsive force that counteracts (and on large scales, overwhelms) the universe's mutual gravitational attraction.

In the 1950s and 1960s there was a rival model, called the steady state theory. The basic assumption of steady state was a perfect cosmological principle, applying to time as well as position. The steady state theory stated that the universe must have the same large-scale properties at all times; it cannot evolve, but must remain uniform. Since the universe is seen to be expanding, which would spread the matter out thinner as time goes on, steady state suggested that new matter must be created to maintain the constant density. In the steady state theory galaxies are formed, they live and die, and new ones come along to take their places at a rate that keeps the average density of matter constant.

When astronomers observe an object at a great distance, they are seeing it as it looked long ago, because it takes time for light to travel. A galaxy viewed at a distance of a billion light-years is seen as it was a billion years ago. Distant galaxies do seem to be different from nearby galaxies. They seem closer together than nearby ones, contrary to steady state contentions but consistent with the view that the universe had a greater density in the past. Also, a faint glow of radiation has been discovered coming uniformly from all directions. Calculations show that this could be radiation left over from the big bang.

age footstock/SuperStock

Stonehenge is an ancient monument in England that includes a circular setting of massive stones. They were precisely aligned in relation to the rising of the sun on the summer solstice.

THE HISTORY OF ASTRONOMY

The ruins of many ancient structures indicate that their builders observed the motions of the sun, moon, and other celestial bodies. The most famous of these is probably England's Stonehenge, which was built between about 3100 and 1550 BC. Some of the monument's large stones were aligned in relationship to the position of the rising sun on the summer solstice. Several hundreds of other ancient structures showing astronomical alignment also have been found in Europe, Egypt, and the Americas.

In many early civilizations, astronomy was sufficiently advanced that reliable calendars had been developed. In ancient Egypt astronomer-priests were responsible for anticipating the season of the annual flooding of the Nile River. The Maya, who lived in what is now central Mexico, developed a complicated calendar system about 2,000 years ago. The Dresden Codex, a Mayan text from the 1st millennium AD, contains exceptionally accurate astronomical calculations, including tables predicting eclipses and the movements of Venus.

In China, a calendar had been developed by the 14th century BC. In about 350 BC a Chinese astronomer, Shih Shen, drew up what may be the earliest star catalog, listing about 800 stars. Chinese records mention comets, meteors, large sunspots, and novas.

The early Greek astronomers knew many of the geometric relationships of the heavenly bodies. Some, including Aristotle, thought Earth was a sphere. Eratosthenes, born in about 276 BC, demonstrated its circumference. Hipparchus, who lived around 140 BC, was a prolific and talented astronomer. Among many other accomplishments, he classified stars according to apparent brightness, estimated the size and distance of the moon, found a way to predict eclipses, and calculated the length of the year to within 6 1/2 minutes.

The most influential ancient astronomer historically was Ptolemy (Claudius Ptolemaeus) of Alexandria, who lived in about AD 140. His geometric scheme predicted the motions of the planets. In his view, Earth occupied the center of the universe. His theory approximating the true motions of the celestial bodies was held steadfastly until the end of the Middle Ages.

In medieval times Western astronomy did not progress. During those centuries Hindu and Arab astronomers kept the science alive. The records of the Arab astronomers and their translations of Greek astronomical treatises were the foundation of the later upsurge in Western astronomy.

In 1543, the year of Copernicus' death, came the publication of his theory that Earth and the other planets revolved around the sun. His suggestion contradicted all the authorities of the time and caused great controversy. Galileo supported Copernicus' theory with his observations that other celestial bodies, the satellites of Jupiter, clearly did not circle Earth.

The great Danish astronomer Tycho Brahe rejected Copernicus' theory. Yet his data on planetary positions were later used to support that theory. When Tycho died, his assistant, Johannes Kepler, analyzed Tycho's data and developed the laws of planetary motion. In 1687 Newton's law of gravitation and laws of motion explained Kepler's laws.

Meanwhile, the instruments available to astronomers were growing more sophisticated. Beginning with Galileo, the telescope was used to reveal many hitherto invisible phenomena, such as the revolution of satellites about other planets.

The development of the spectroscope in the early 1800s was a major step forward in the development of astronomical instruments. Later, photography became an invaluable aid to astronomers. They could study photographs at leisure and make microscopic measurements on them. Even more recent instrumental developments—including radar, telescopes that detect electromagnetic radiation other than visible light, and space probes and manned spaceflights—have helped answer old questions and have opened astronomers' eyes to new problems.

FURTHER RESOURCES

Asimov, Isaac, and Hantula, Richard. Our Solar System, rev. and updated ed. (Prometheus Books, 2004).
Asimov, Isaac, and Giraud, Robert. The Future in Space (Gareth Stevens, 1993).
Beatty, J.K. Exploring the Solar System: Other Worlds (National Geographic, 2001).
Bell, Jim, and Mitton, Jacqueline, eds. Asteroid Rendezvous: NEAR Shoemaker's Adventures at Eros (Cambridge Univ. Press, 2002).
Benton, Julius. Saturn and How To Observe It (Springer, 2005).
Burnell, S. Jocelyn Bell, and others. An Introduction to the Sun and Stars (Open Univ.–Cambridge Univ. Press, 2004).
Burnham, Robert. Great Comets (Cambridge Univ. Press, 2000).
Cattermole, P. J. Building Planet Earth, rev. ed. (Cambridge Univ. Press, 2000).
Chapman, C.R., and Morrison, David. Cosmic Catastrophes (Plenum, 1989).
Comins, N.F. What if the Moon Didn't Exist: Voyages to Earths that Might Have Been (HarperPerennial, 1995).
Croswell, Ken. Ten Worlds: Everything That Orbits the Sun (Boyds Mills, 2006).
Dorminey, Bruce. Distant Wanderers: The Search for Planets Beyond the Solar System (Copernicus, 2002).
Elkins-Tanton, L.T. The Sun, Mercury, and Venus (Chelsea House, 2006).
Fischer, Daniel. Mission Jupiter: The Spectacular Journey of the Galileo Spacecraft (Copernicus, 2001).
Friedlander, M.W. Astronomy: From Stonehenge to Quasars (Prentice, 1985).
Goldsmith, Donald. The Astronomers (St. Martin's, 1991).
Golub, Leon, and Pasachoff, J.M. Nearest Star: The Surprising Science of Our Sun (Harvard Univ. Press, 2002).
Hartmann, W.K. A Traveler's Guide to Mars: The Mysterious Landscapes of the Red Planet (Workman, 2003).
Hubble, E.P. The Realm of the Nebulae (Yale Univ. Press, 1982).
Hunt, G.E., and Moore, Patrick. Atlas of Uranus (Cambridge Univ. Press, 1989).
Illingworth, Valerie, and Clark, J.O.E., eds. The Facts on File Dictionary of Astronomy, 4th ed. (Facts on File, 2000).
Irwin, P.G.J. Giant Planets of Our Solar System: An Introduction (Springer-Praxis, 2006).
James, Nick, and North, Gerald. Observing Comets (Springer, 2003).
Kaler, J.B. The Cambridge Encyclopedia of Stars (Cambridge Univ. Press, 2006).
Knapp, B.J. Rocky Planets (Grolier, 2004).
Koerner, David, and LeVay, Simon. Here Be Dragons: The Scientific Quest for Extraterrestrial Life (Oxford Univ. Press, 2001).
Lang, K.R. Sun, Earth, and Sky, 2nd ed. (Springer, 2006).
Lederman, L.M., and Schramm, D.N. From Quarks to Cosmos: Tools of Discovery, rev. ed. (Scientific American Library, 1995).
Leutwyler, Kristin. The Moons of Jupiter (Norton, 2003).
Lippincott, Kristen. Astronomy, rev. ed. (DK Publishing, 2004).
Lorenz, Ralph, and Mitton, Jacqueline. Lifting Titan's Veil: Exploring the Giant Moon of Saturn (Cambridge Univ. Press, 2002).
Lovett, Laura, and others. Saturn: A New View (Abrams, 2006).
Marvel, Kevin. Astronomy Made Simple (Broadway Books, 2004).
Mathez, E.A., and Webster, J.D. The Earth Machine: The Science of a Dynamic Planet (Columbia Univ. Press, 2004).
Melton Knocke, Melanie. From Blue Moons to Black Holes: A Basic Guide to Astronomy, Outer Space, and Space Exploration (Prometheus Books, 2005).
Menzel, D.H., and Pasachoff, J.M. A Field Guide to the Stars and Planets, 4th ed. (Houghton, 2000).
Miller, Ron. Earth and the Moon: Worlds Beyond (Twenty-First Century Books, 2003).
Miner, E.D., and Wessen, R.R. Neptune: The Planet, Rings, and Satellites (Springer, 2002).
Moore, Patrick. The Guinness Book of Astronomy, 5th ed. (Guinness, 1995).
Moore, Patrick. Venus (Cassell, 2002).
Morris, Richard. Cosmic Questions: Galactic Halos, Cold Dark Matter, and the End of Time (Wiley, 1998).
Mortin, Oliver. Mapping Mars: Science, Imagination, and the Birth of a World (Picador, 2003).
Odenwald, S.F. Back to the Astronomy Café: More Questions and Answers About the Cosmos from "Ask the Astronomer" (Westview, 2003).
Pasachoff, J.M. Astronomy: From the Earth to the Universe, 6th ed. (Brooks/Cole–Thomson Learning, 2002).
Pasachoff, J.M., and others. Peterson First Guide to the Solar System, updated ed. (Houghton, 1998).
Peebles, Curtis. Asteroids: A History (Smithsonian, 2001).
Phillips, K.J.H. Guide to the Sun (Cambridge Univ. Press, 1995).
Ride, Sally, and O'Shaughnessy, T.E. Exploring Our Solar System (Crown, 2003).
Ridpath, Ian. The Facts on File Stars & Planets Atlas, 4th ed. (Facts on File, 2004).
Rothery, D.A. Satellites of the Outer Planets: Worlds in Their Own Right, 2nd ed. (Oxford Univ. Press, 1999).
Scholastic Atlas of Space (Scholastic Refernce, 2004).
Skurzynski, Gloria. Are We Alone?: Scientists Search for Life in Space (National Geographic, 2004).
Sparrow, Giles, and Kerrod, Robin. The Way the Universe Works, 1st American ed. (DK Publishing, 2002).
Squyres, S.W. Roving Mars: Spirit, Opportunity, and the Exploration of the Red Planet (Hyperion, 2005).
Standage, Tom. The Neptune File: A Story of Astronomical Rivalry and the Pioneers of Planet Hunting (Walker, 2000).
Stern, Alan, and Mitton, Jacqueline. Pluto and Charon: Ice Worlds on the Ragged Edge of the Solar System, 2nd ed., rev. and updated (Wiley-VCH, 2005).
Strom, R.G., and Sprague, A.L. Exploring Mercury: The Iron Planet (Springer, 2003).
Sumners, Carolyn, and Allen, Carlton. Cosmic Pinball: The Science of Comets, Meteors, and Asteroids (McGraw Hill, 2000).
Tocci, Salvatore. A Look at Uranus (Franklin Watts, 2003).
Ulivi, Paolo, and Harland, D.M. Lunar Exploration: Human Pioneers and Robotic Surveyors (Springer-Verlag, 2004).
Villard, Ray, and Cook, Lynette. Infinite Worlds: An Illustrated Voyage to Planets Beyond Our Sun (Univ. of Calif. Press, 2005).
Weintraub, D.A. Is Pluto a Planet?: A Historical Journey Through the Solar System (Princeton Univ. Press, 2006).
Wilson, D.A. Star Track: Plotting the Locations of the Stars Within 30 Light-Years of Earth (Lorien House, 1994).
Zimmerman, Robert. The Chronological Encyclopedia of Discoveries in Space (Oryx, 2000).

INDEX

A

B

C

D

E

F

G

H

I

J

K

L

M

Page numbers in **bold** indicate main subject references; page numbers in *italics* indicate illustrations.

N

O

P

Q

R

S

T

U

V

W

X

Y

Page numbers in **bold** indicate main subject references; page numbers in *italics* indicate illustrations.